SIGNAL TRANSDUCTION IN CANCER

SIGNAL TRANSDUCTION IN CANCER

Mohamed A. Selmy

Delve
PUBLISHING
www.delvepublishing.com

Signal Transduction in Cancer

Mohamed A. Selmy

Delve Publishing

2010 Winston Park Drive,

2nd Floor

Oakville, ON L6H 5R7

Canada

www.delvepublishing.com

Tel: 001-289-291-7705
001-905-616-2116
Fax: 001-289-291-7601

Email: orders@arclereducation.com

ISBN: 978-1-77361-484-7 (Hardcover)

ABOUT THE AUTHOR

Mohamed Selmy is a physician scientist and lecturer of Medical Biochemistry and Molecular Biology. He received his medical degree in 2004 from the faculty of medicine Suez Canal University in Egypt. He received his scientific training through many fellowships in United States, Spain and Italy. He received his PhD in Medical Biochemistry and Cancer biology in 2017 from the faculty of medicine Suez Canal University in Egypt. His fields of expertise comprises; cancer biology, cell signaling and molecular medicine. His work is continuously published in international peer reviewed journals.

TABLE OF CONTENTS

LIST OF FIGURES

LIST OF TABLES

LIST OF ABBREVIATIONS

AIF	Apoptosis-inducing factors
AMPK	AMP-dependent kinase
ANG	Angiopoietins
APAF	Apoptotic peptidase-activating factor
ATRA	All-trans retinoic acid
CARDs	Caspase recruitment domains
CASP3	Caspase 3
CBL	Casitas B-lineage lymphoma
CDK	Cyclin dependent kinase
cGMP	cyclic guanosine monophosphate
DAG	Diacylglycerol
DLL	Delta-like ligands
ECM	Extracellular matrix
EGF	Epidermal growth factor
EMT	Epithelial-to-mesenchymal transition
ERK	Extracellular signal-regulated kinases
FGF	Fibroblastic Growth Factors
FLT	FMS-like tyrosine kinase
GAP	GTPase- activating protein
GEF	GDP–GTP exchange factor
GIST	Gastrointestinal stromal tumors
GPCR	G-Protein-Coupled Receptors
HDAC	Histone deacetylase
HGF	Hepatocyte Growth Factor
HIF	Hypoxia-inducible factor
HSP	Heat shock protein

IGF	Insulin-Like Growth Factors
IL	Interleukin
IRS	Insulin receptor substrates
LKB1	Liver kinase B1
MCSF	Macrophage colony-stimulating factor
MMPs	Matrix metalloproteinases
mTOR	mechanistic target of rapamycin
NO	Nitric Oxide
NPP	Natriuretic peptides
PDGF	Platelet-Derived Growth Factors
PI3K	Phosphatidylinositol 3-kinases
PIP	Phosphatidylinositol 4,5-bisphosphate
PLC	Phospholipase C
PTB	Phosphotyrosine-binding domains
ROS	Reactive Oxygen Species
SH2	SRC homology domains 2
TACE	TNF-alpha-converting enzyme
TGF α	Transforming growth factor α
TSC	Tuberous sclerosis complex
VEGF	Vascular Endothelial Growth Factor
VHL	von Hippel–Lindau

PREFACE

There is a fundamental necessity for oncologists to identify cell–cell signaling pathways, to comprehend their interactions, their role in cell proliferation, motility, survival and their contribution in cancer. The clinical oncology practice has turned to be dependent upon the awareness of cancer biology. The understanding of the pathways and their downstream signals would restrain the toxic and expensive treatments to half of the patients suffering from certain metastatic cancers.

This book aims at illustrating the signaling pathways to readers involved in the oncology practice. The key cell signaling pathways are presented in independent chapters, each of them illustrating a "signaling system" characterized by the intimate pairing of a receptor type and its related ligands. Despite the complexity of these pathways, I have tried to make it palatable and simple as possible as I can. In order to provide a complete understanding, the general mechanisms governing DNA repair, gene regulation, epigenetics and protein expression have been illustrated, together with the changes of these mechanisms that share in the process of oncogenesis. As the crosstalks and interactions between signaling pathways are so diverse and crucial, a note was given at the each chapter illustrating the interaction among pathways. In addition, each chapter was tailed by a short review of the oncogenic changes of the pathway studied and by a short appraisal of the pharmacological targets that may be put in concern for therapeutic agents. The references, provided at the end of each chapter, were limited to selected updated reviews about the topic of the chapter. In its final format, this book represents an informative summary of many books covering the topic of oncological cell signaling; especially the book "Textbook of Cell Signalling in Cancer" by Jacques Robert which we shadowed its format. I have tried to give a useful summary of these books for a quicker and focused revision.

There are numerous options for the revision of cell signaling in oncology. One can start from the demonstration of the players that interfere in all pathways: ligands, ligands, protein kinases, transcription factors and small G-proteins and define their numerous roles. Another way is to provide a "horizontal" combination level: plasma membrane, cytoplasm, mitochondria and nucleus. It is also possible, as illustrated here in this book, to shape a "vertical" level of interaction, presenting each pathway by the type of receptor complicated and illustrating the available linked data: ligands, receptor activation, the information transduction and execution of effectors.

In relation, despite the density and the numerous connectivity of signaling pathways, there are homogenous receptor clusters, able to deduce the signals brought by ligands belonging to certain families: this book is framed around this ligand–receptor alignment. This way permits a more simplified illustration of signaling modules but it may retain some difficulties at the level of the "shared final pathways" of cell signaling, it would have been conceivable to write special chapters on final transcription factors, but it would be beyond the scope of this book.

I hope that this book will realize its goal in yielding the necessary information regarding the oncological signaling pathways in cancer. Hopefully, these summary notes will keep the reader safe in the jungle of the pathways, yet giving him a concise review of structures, mechanisms and therapeutic targeting of these pathways.

INTRODUCTION: BASICS AND TRACKS OF CELL SIGNALING

Cell communication is vital for multicellular organisms: cells must inevitably give-and-take the information compulsory for organizing their activities. The main lines of cell communication are generally universal: signaling molecules are released by a known cell and identified by another one, which in turn stimulates a transduction pathway, allowing an effector system capable of achieving the matching tasks. The attainment of this information transmission necessitates from the source cell the encrypting of the communication message in a way that could be appropriately understood, and from the cell in receipt of the message to the agreeing decoding structures (Fig. 1). The diversity of signal transduction systems is extensive, at the level of both signal reception and task implementation. Nevertheless, it is definitely probable to recognize general outlines and shared structures of organization.

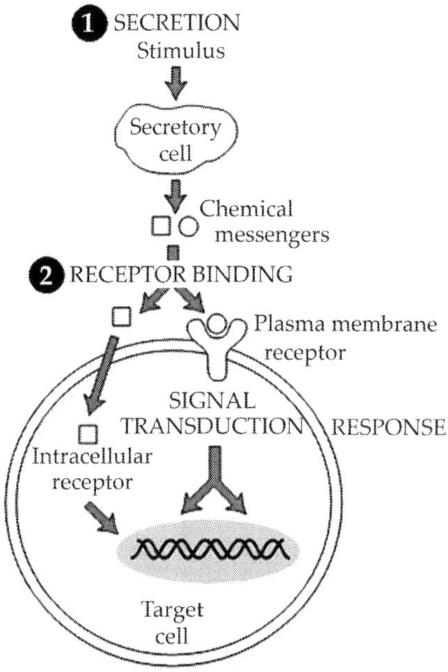

Figure. 1: The mechanism of signaling from the message to the effector (Lehninger, 2004).

At the level of signal response, the mechanisms are generally reliant on the chemical structure of the messengers:

- Hydrophilic messengers (proteins, peptides, amino acids and derivatives) cannot go in the cells as they cannot simply cross the membranes; a membrane receptor is a prerequisite to attain and comprehend the message and to transmit the information afterwards.
- Lipophilic messengers (steroids, fatty acids and derivatives) and the simple compounds (oxygen, nitric oxide) are capable of diffusing inside cell membranes and to affect directly their corresponding intracellular targets, in the cytoplasm or the nucleus.
- Ionic compounds (Na^+, K^+, Cl^- and Ca^{2+}) are capable of induction of opening or closing of membrane channels permitting the production of transmembrane currents which are employed for the transmission of nervous impulses but are also accountable for numerous intracellular actions. The transmission of the signals

attained by the receptors trails many different processes but the general mechanisms are not plentiful, the main mechanisms are:

- Enrollment of adapter proteins capable of interacting with other proteins and inducing conformational changes, and subsequently protein activity.
- The phosphorylation and dephosphorylation cascades, stimulated by kinases and phosphatases, which adjust the protein three-dimensional structures.
- The activation of small G-protein, through exchange and hydrolysis processes of guanyl nucleotides, which also stimulate alterations in the conformation of protein.
- Production of second intracellular messengers, which transmit the information, carried to the membrane by the extracellular first messengers.
- In addition, the effectors are also various, but again it is conceivable to group them together in a few assorts:
- Transcriptional factors which order target genes transcription; these are the most classical effectors and the most frequently employed downstream the signal transduction pathways.
- Translational regulators executing on protein synthesis, which are directly integrated in certain signaling pathways.
- Proteins of the cytoskeleton or the extracellular matrix, which regulate the cellular adhesion, motility and migration.
- Ionic channels involved especially (but not exclusively) in synaptic transmission.

The cell, consequently, has a 'toolbox' from which it can draw the sufficient tools to comprehend the information established and perform the given orders.

On the other hand, cell signaling may also work at highly flexible distances: the first signaling system that has been recognized is the endocrine system, in which the endocrine gland secretes hormones targeting distant cells and organs. The molecules produced in a certain cell could be delivered to the neighboring cells, which is termed 'paracrine' signaling; also the term 'juxtacrine' is given for the signal transmitted between jointed cells; and ultimately the term 'autocrine' is established when the molecules, after transiting in the extracellular space, execute their effect on the same cell. In addition to the intercellular signaling, an intracellular signaling mechanism also exists, and both are integrated and barely separable. This way of communication is particularly achieved by the second messengers mentioned earlier, and also by a count of proteins, which are produced in reaction to all assorts of demands.

For a clustering illustration, the information transported from one cell to another can be assorted into six bulky types of commands to be performed, which represent opposite pairs: to survive or to die, to proliferate or to differentiate and to stay attached or to migrate; one can discriminate, as 'survival pathways,' 'proliferation pathways' and 'motility pathways' and it is quite often, the same signals can convey these dissimilar commands: proliferation signals can turn to be differentiation signals in certain situations and survival signals can become death signals in certain cell types, depending on the receptor of the cell attaining the message.

Therefore, when studying cell signaling, especially in cancer, the simplified linear model drawn by the endocrinologists (one ligand, one receptor, one message, one physiological action) – must be restrained. In fact, here the same ligand can bind numerous receptors (which may have contradictory roles); and the same receptor can accept information from diverse ligands; the intracellular signals transduced after receptor stimulation are generally numerous and the physiological effects are flexible from one tissue to another. The unique apparatus of each cell or tissue in terms of receptors and transduction systems is accountable for the final behavior of the cells which attain a message; this illustrate why the same message can stimulate cell differentiation in a given tissue and cell division in another. The actions of the signals exchanged between cells are diverse and depends upon the cellular milieu.

Cell Signaling and Cancer

We will not designate here the mechanisms complicated in oncogenesis but show some features of malignant cells to reveal how the molecular modifications taking place in cell signaling pathways establish the foundations of oncogenesis. The cancer cell is generally genetically unstable, and is therefore capable of exploring the possibilities in the whole genome, to choose any proliferative or migratory benefit, and to convey it to its descent. The proliferative track is crucial as any tumor is a neoformation necessitating active cell reproductive capability; migratory pathway is also indispensable as the malignancy potential of cancers lies in their ability to disseminate in the organism. All signaling pathways integrated in cell survival and death, cell proliferation and differentiation and in cell adhesion and migration, may concrete the way to oncogenic changes.

In that context, cancer looks as a disease of cell signaling. In a general review elaborated by Hanahan and Weinberg, the mechanisms of oncogenesis were classified in six large families (1, 2); their contribution, with some revisions, is still valid (Fig. 2).

- Self-sufficiency in growth signals
- Insensitivity to anti-growth signals
- Apoptosis evasion
- Limitless replicative potential
- Sustained angiogenesis
- Tissue invasion and metastasis

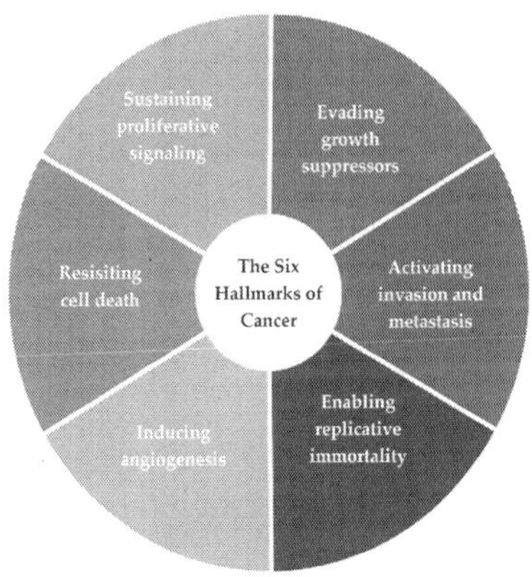

Figure 2. The 'hallmarks' of cancer (Hanahan and Weinberg, 2000).

One can supplement another general mechanism which paves the way to all others: genomic instability. For all mechanisms, it is always likely to recognize a modification in information transfer initiating from a mutation in a protein integrated in signal transduction, permitting a disease of cell signaling.

The cancer cell way to divide, grow and disseminate, definitely represents the best prototype to study Darwinian evolution at the cellular level. It would be so simple, if the cancer cell would only modify the chief mechanisms that control epithelial cell proliferation and migration; but in adding, it is capable of seeking unanticipated pathways integrated in cell polarity, inflammation, intercellular junctions, axonal migration and many other specialized processes that appeared restricted to a precise tissue or to a specific step of embryonic development. To comprehend the hijacking of cell signaling pathways performed by cancer cells, one should not be restricted to study understandable processes, but look for the uncommon, unrecognized and even exceptional characters.

The comprehension of oncogenetic mechanisms and of signaling pathways modification in cancers has allowed for the emergence of the notion of targeted therapies. As an alternative to the classical chemotherapy in which drugs target only the effectors of cell multiplication, normal or neoplastic, this novel approach aims at targeting the actual mechanisms of oncogenesis and cancer development, and should have, accordingly, a higher degree of selectivity to cancer cells. In these targeted therapies, the targets are the proteins integrated in the regulation of cell proliferation, migration and survival, and no longer are responsible for the accomplishment of these programs; in other words, the ground of the abnormality producing a cancer, and not the effect subsequent to this abnormality. These therapies have been released when the comprehension of cancer biology has reached a reasonable level and when the tools permitting the interference with malignant transformation have become attainable.

REFERENCES

1. Lehninger, A. L., Nelson, D. L. and Cox, M. M. (2004). *Lehninger principles of biochemistry* (4th ed.). New York: Worth Publishers.
2. Hanahan, D, Weinberg, RA (2000). The hallmarks of cancer. *Cell*, *100*, 57-70.
3. Hanahan, D, Weinberg, RA (2011). Hallmarks of cancer: the next generation. *Cell*, *144*, 646-74.

1
CHAPTER

GROWTH FACTORS AND TYROSINE KINASE RECEPTORS

1.1 GENERAL CHARACTERISTICS OF THE GF–TKR INTERRELATIONSHIP

1.1.1. Introduction

The different growth factors (GFs) and tyrosine kinase receptors (TKRs) can be clustered in groups of homologous proteins: for example, the factors of the epidermal growth factor (EGF) family cooperate with the TKRs of the epidermal growth factor receptor (EGFR) family, and it is the same for other groups. The terms 'fibroblastic' or epidermal GFs have additional tasks to execute with the settings of their recognition than with the specificity of their acknowledged target. GFs are polypeptides conveying mitogenic orders, that is, orders capable of triggering mitosis. They are released by various cell sorts, not organized in tailored glands, and act mostly in the near neighborhood of the original cell. By similarity (and antagonism) to the endocrine system, consisted of circulating hormones in the blood, GFs are known to be paracrine factors. They may also work as juxtacrine factors and as autocrine factors.

Table 1 shows the 20 sorts of TKRs and the related GFs. All TKRs show a single transmembrane domain; a homologous intracellular domain possessing the catalytic center, of limited structural diversity; and an extracellular part which is exceedingly variable from one class to another.

EGFR against which numerous drugs have been marketed; numerous other TKRs have been taken in account for targeting and suitable drugs are under progress.

1.2 THE PARADIGMATIC MODEL OF THE EGF (ERBB) FAMILY

1.2.1 Growth Factors and Their Receptors

Eleven GFs of the EGF cluster have been recognized: EGF itself, epiregulin (EREG), heparin-binding EGF (HBEGF), β-cellulin (BTC), TGFα (transforming growth factor α), amphiregulin (AREG), epigen or epithelial mitogen (EPGN) and four neuregulins (NRG1-4). These factors exert a key role in the proliferation capability of several tissues. EGF is a small soluble molecule of 53 amino acids, with three intramolecular disulphide bridges (Figure. 4a). It originates from a voluminous transmembrane protein of 1,207 amino acids comprising a series of analogous EGF motifs whose discharge by cleavage in the extracellular space delivers soluble factors. The membrane insertion of native EGF allows its action on adjoining cells, explaining, therefore, a juxtacrine activity. The other factors of the EGF cluster are also synthesized as transmembrane precursors but comprise only one EGF extracellular motif. The cleavage of the transmembrane precursors into soluble GFs is achieved by metalloproteinases such as ADAM17 (a disintegrin and metalloproteinase) or TNF-alpha-converting enzyme (Figure. 4b).

Table 1: Tyrosine kinase inhibitors (Robert, 2010)

TM name	Code number	Pharmaceutical laboratory	Target	Present status
Gefitinib	ZD1839	AstraZeneca	EGFR (reversible)	Phase III (Iressa[e])
Erlotinib	OS-774	Roche	EGFR (reversible)	Approved (Tarce-va[e])
Lapatinib	GW572016	GSK	EGFR, ERBB2	Approved (Tyverbe)
Canertinib	CI-1033	Pfizer	Pan-ERBB (irreversible)	Phase II

Afatinib	BIBW-2992	Boehringer Inge!helm	EGFR, ERBB2 (reversible)	Phase III
Dacomitinib	PF-00299804	Pfizer	Pan-ERBB (irreversible)	Phase III
	HM781-36B	Hanmi	Pan-ERBB (irreversible)	Phase I
Neratinib	HKI-272	Wyeth	EGFR,, ERBB2, (irreversible)	Phase II
Pelitinib	EKB-569	Wyeth	EGFR, ERBB2, (irreversible)	Phase II
Vandetanib	ZD6474	AstraZeneca	EGFR, FGFR1, VEGFR2. RET	Approved (Zactima[e] or Caprelsae)
Sunitinib	SU-11248	Pfizer	VEGFR, PDGFR, KIT, FLT3, FGFR I	Approved (Sutent[e])
Sorafenib	BAY-439006	Bayer	B-RAF,, RET, VEGFR, FGFRI. FLT3	Approved (Nexavar[e])
Vatalanib	PTK787/ ZK222584	Novartis	VEGFR, KIT	Phase III
Axitinib	AG-013736	Pfizer	VEGFR, PDGFR, KIT	ATU (Inlytae)
Semaxanib	SU-5416	Pharmacia	KIT, FLT3, VEGFR2	Stopped
Cediranib	AZD-2171	AstraZeneca	VEGFR, PDGFR	Phase III
Motesanib	AMG-706	Amgen	VEGFR, PDGFR, RET, KIT, FLT3	Phase II
Pazopanib	GW-786034	GSK	VEGFR, PDGFR	Phase III
Imatinib	STI-571	Novartis	BCR-ABL, KIT. PDGFR	Approved (Glivec[e])
Nilotinib	AMN-107	Novartis	BCR-ABL, KIT, PDGFR, RET	Approved (Tasigna[e])
Dasatinib	BMS-354825	BMS	BCR-ABL, KIT, PDGFR	AMM (Sprycel[e])
Masitinib		AB Science	FGFR3, PDGFR, KIT	Phase III
Lestaurtinib	CEP-701	Cephalon	FLT3, NTRK, JAK2	Phase III
Crizotinib	PF-02341066	Pfizer	ALK, MET	Approved (Xalkori[e])
Regorafenib	BAY-73-4506	Bayer	VEGFR, KIT, RET	Approved (Stivargae)

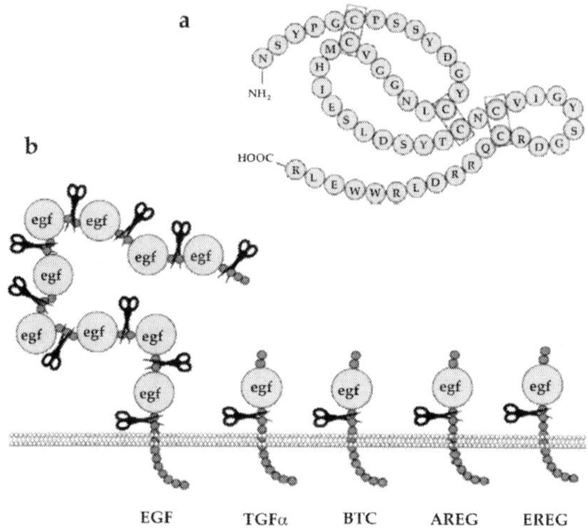

Figure 4. Epidermal growth factor (EGF) family growth factors (GFs) (Robert, 2010).

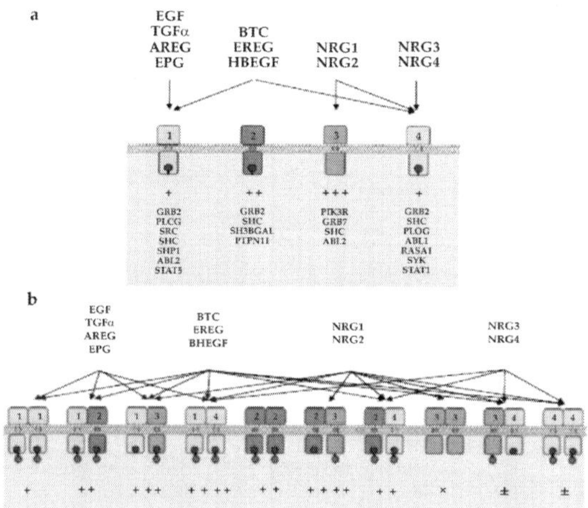

Figure 5. GFs of the ERBB family (Robert, 2010).

These 11 GFs are recognized by four dissimilar ERBB (class I) receptors (EGFR or ERBB1, ERBB2 or HER2, ERBB3, ERBB4). There is a composite combinatorial outline of interaction between the 11 GFs and the four receptors capable of detecting them and activate a signaling pathway. This interaction results in the homo- or heterodimerization of two molecules of the receptor. EGF, TGFα, amphiregulin and epigen only recognize and bind EGFR itself; HBEGF, epiregulin and β-cellulin recognize and bind EGFR and ERBB4; neuregulins 1 and 2 recognize and bind ERBB3 and ERBB4, and neuregulins 3 and 4 ERBB4. There are two particularities: the ERBB2 receptor bears no ligand (deaf receptor) and cannot experience homodimerization in normal circumstances; the ERBB3 receptor bears no tyrosine kinase activity (dumb receptor) and its homodimers are inactive and cannot execute a signal.

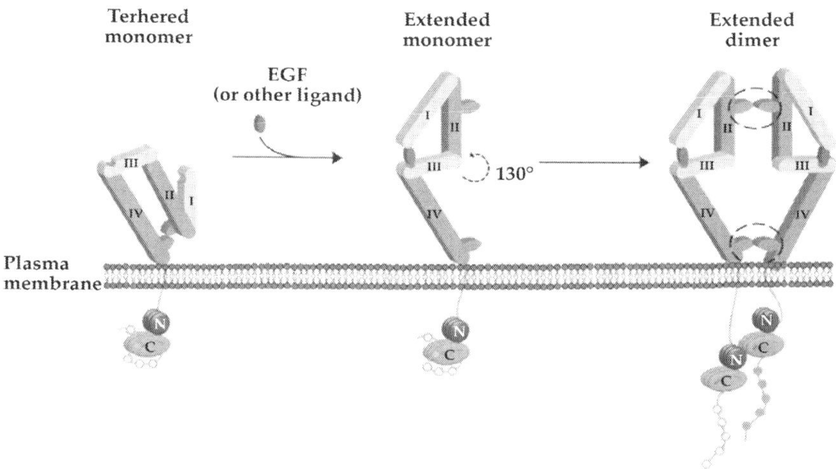

Figure. 6 EGFR dimerization; in the absence of ligand, and in the presence of a ligand such as EGF (Nevo, 2010).

1.2.2 ERBB Receptors Activation

EGFR dimerization is allowed by the exposing of a binding site between two molecules of the receptor, this unveiling being brought by the binding of each molecule of ligand to a molecule of the receptor. At the inactive condition, the receptor adopts a 'tethered' conformation, not permitting the interaction between two molecules of the receptor (Figure. 6). The binding of EGF, at the level of the extracellular domains 1 and 3, permits the conversion to an 'extended' conformation, which exposes a binding site for another molecule of receptor possessing identically bound a molecule of ligand. The ERBB2 receptor has no ligand as it is always under extended conformation and, hence, can be dimerized with other receptors without previous activation.

The dimerization step permits the contact between two intracellular catalytic domains. Their interaction brings the opening of the active center of the enzyme and permits ATP to access this center, therefore called the ATP pocket; the catalytic tyrosine kinase activity is then operative. The alignment of the two intracellular domains is asymmetric, one of the catalytic domains being capable of phosphorylating the tyrosine residues of the C-terminal end of the second one. The phosphotyrosine residues may then be recognized and bound by a great diversity of proteins possessing an SH2 (SRC homology-2) or a PTB (phosphotyrosine-binding) domain. The two types of signaling illustrated above can be then utilized, firstly through the activation of kinases or phosphatases such as SRC or SHP1, which are phosphorylated and activated by the tyrosine kinase activity of the receptor, and furthermore through adapter proteins such as GRB2 or p85α (gene PIK3R1). The receptor phosphotyrosine residues that are recognized by each SH2- or PTB-domain-comprising protein have been recognized: for example, SRC recognizes and binds with high affinity the EGFR phosphotyrosine residue at position Tyr 974, GRB2 the residues Tyr 1148 and Tyr 1173 and so forth. Additionally, the numerous ligands that activate EGFR encourage the phosphorylation of distinct tyrosine residues: therefore, TGFα or AREG may induce specific effects, dissimilar from those induced by EGF.

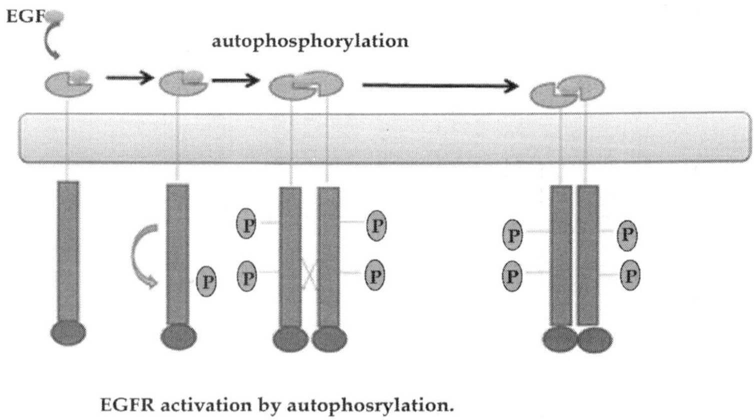

Figure 7. EGF receptor autophosphorylation. (a) Before receptor activation, (b) Upon receptor activation (Pecorino, 2008).

The numerous ERBB receptors are activated by their ligands and activate in turn the downstream effectors agreeing to the same general mechanisms, concerning homo or heterodimerization and autophosphorylation. In addition to the numerous combinations previously mentioned (homodimers and heterodimers) and the numerous GFs that encourage their activation.

ERBB receptors are internalized by endocytosis and are consequently either recycled or deactivated by transfer into the lysosome. Ligand binding speedup internalization and reduces thus the accessibility of the receptor at the surface of the cell. Internalization takes place within clathrin-containing vesicles and can be tailed, after transfer into endosomes, by ubiquitinylation by an E3 ubiquitin ligase called Casitas B-lineage lymphoma, which shows a binding domain for phosphotyrosine residues.

1.2.3 Oncogenic Changes

The amplification of the EGFR and ERBB2 genes are oncogenic happenings; for EGFR, this amplification takes place in numerous types of epithelial cancers, but its significance in oncogenesis is still debated; for ERBB2, an amplification takes place in 15–25 % of breast cancers and represents a distinguishing oncogenic event, outlining the ERBB2-positive cancers and permitting specific targeted therapies. ERBB2 amplification also takes place in gastric cancers and some other malignancies. The mechanism of gene amplification rests incompletely unstated.

Two types of oncogenic mutations can be established in the EGFR gene. The first one employs the extracellular part of the protein, a part of which is deleted, resulting in a permanent activation of the receptor in the absence of ligand. The VIII mutation is the most common one and is present particularly in glioblastoma. The second type implements the intracellular catalytic site of the receptor (tyrosine kinase activity). These mutations have been detected basically in a small proportion (5–10%) of non-small cell lung cancers and permit the permanent accessibility of the catalytic site to ATP: they allow the autophosphorylation of the tyrosine residues in the absence of a mitogenic signal and of receptor dimerization. These mutations existed on exon 19 (various deletions in codons 747–751, 45%), exon 18 (G719X, 5% of activating mutations), exon 20 (rare insertions) and exon 21 (L858R, 50%, the most frequent one). Since they maintain the ATP-binding domain open, they are also accountable for the sensitivity of tumor cells to inhibitors of the tyrosine kinase activity of the receptor, being therefore instantaneously oncogenic and drug-sensitizing mutations. These mutations must be obviously notable in relation to other mutations, also taking place close to the catalytic domain of EGFR, which conversely hinder the access of the TKI to the active site and allow resistance to these drugs; the most common of these mutations is the T790G mutation.

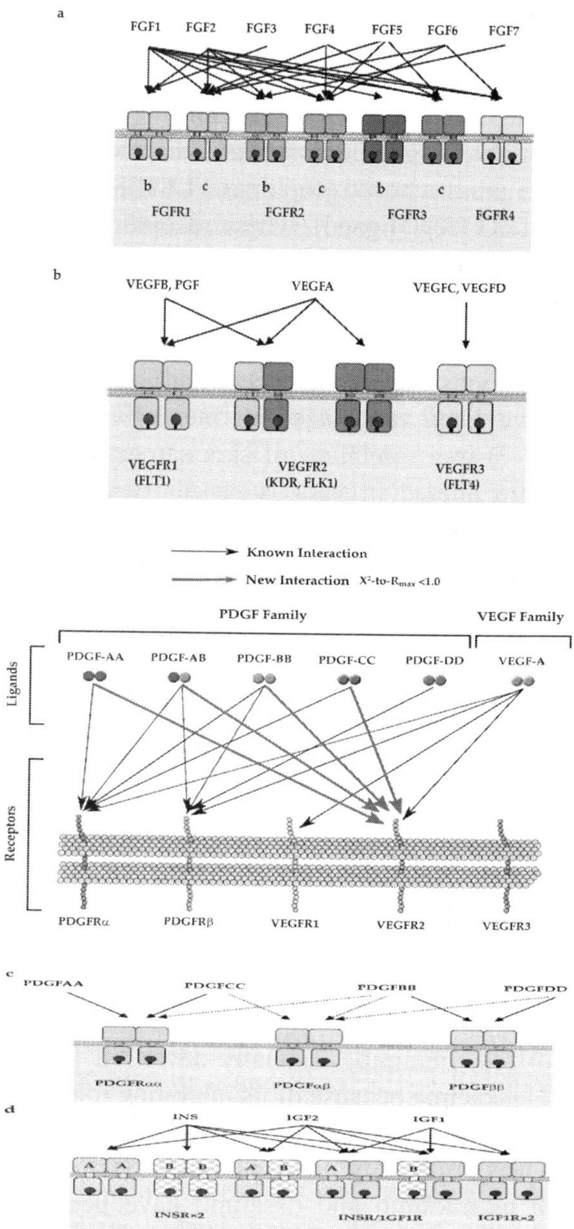

Figure. 8 The pattern of the interaction between GFs and their cognate receptors is presented for some members of the fibroblastic GF family (a) and for the members of vascular endothelial GF (b), PDGF (c) and insulin-like GF (d) families (Robert, 2010).

1.3.2 Insulin-Like GF and Their Receptors

Insulin is a known pancreatic hormone and shares numerous characteristics with GFs, especially the notion that it detects and binds a TKR. Insulin-like GFs (IGF1 and IGF2) or somatomedins, analogues of insulin, are produced essentially by the liver and also execute its actions through the stimulation of a TRK. The insulin receptor (INSR) presents in two tissue-specific isoforms, A and B, due to an alternative splicing. Only insulin can stimulate the homodimerization of INSR-B, expressed only in specific tissues, which results in the traditional effects of this hormone on carbohydrate metabolism. Insulin and IGF2 can also stimulate other INSR dimerization processes, involving the A isoform, expressed in a variety of epithelial tissues. IGF1 and IGF2 can stimulate the homodimerization of another ubiquitous receptor, IGF1R, as well as the heterodimerization of INSR (A or B) with IGF1R, which is also acquired by insulin itself (Figure. 8). INSR and IGF1R show a classical structure among TKRs, with two polypeptide chains linked through disulphide bonds, one chain being mostly intracellular and the other is only extracellular. There is also a false receptor, IGF2R, which can bind IGF2 without being capable of transmitting a signal and work, consequently, as an inhibitor of this signaling pathway.

The activation of the target proteins takes place through the initial phosphorylation of docking proteins, the INSR substrates, which are, in consequence, detected by the SH2 domain of the regulatory subunit of PI3 kinase; the PI3 kinase pathway is the chief pathway activated downstream insulin and IGF binding to their related receptors.

Regarding oncogenesis, it has been proposed that high circulating levels of IGF1 could be linked to the risk of cancer. INSR (isoform A) and IGF1R are commonly overexpressed in cancer cells, typically in the lack of gene amplification; no oncogenic mutations have yet been recognized. IGF2R would relatively exert a tumor-suppressor function, through binding to IGF2 without signal production, though no loss-of-function mutation has been identified.

Currently, monoclonal antibodies against the GFs IGF1 and IGF2 are in preclinical progress; additionally, numerous monoclonal antibodies against IGF1R, such as figitumumab or dalotuzumab, have gained entry to clinical trials. Tyrosine kinase inhibitors active against IGF1R have also been recognized; targeting the receptor with monoclonal antibodies seems more favorable than with TKIs as they possess a high specificity toward the receptor, while TKIs may impede the INSR with likely unfavorable metabolic side effects.

1.3.3. Fibroblastic GF and Their Receptors

There are 18 GFs of this family, numbered after fibroblastic GF (FGF1) (aFGF, FGFα or acidic FGF) and FGF2 (bFGF, FGFβ or basic FGF). FGFs exert various roles in the development of various tissue types; particularly, FGF2 is an angiogenic factor. FGFs are detected by four discrete receptors, FGFR1 to FGFR4, with the exception of a group linked to FGFs named FHF (FGF homologous factors), which possess no receptor-binding site. FGF activity is frequently dependent upon the existence of a membrane glycosaminoglycan, heparan sulfate, which exerts a coreceptor effect for the FGFs. FGFR1, FGFR2 and FGFR3 each present in two isoforms, b and c, as a result of the alternative splicing and expressed either in epithelial tissues or in mesenchymal (conjunctive) tissues, with dissimilar affinities to their substrates. As for EGFR, FGF receptor dimerization is needed for activation. However, it appears that another mode of dimerization would work; it would comprise the presence of two binding sites for the growth factor, one with high and the other with low affinity, and secondly the presence of a binding site of the receptor with heparin sulfate (Figure. 8). Finally, the activation of the effector substrates of the tyrosine kinase activity of FGFRs frequently needs the tyrosine phosphorylation of intermediate docking proteins called FRS (FGF receptor substrates).

Regarding oncogenesis, FGF3, the homologue of the murine oncogene Int2, is amplified in numerous sorts of human tumors. Oncogenic changes of FGFRs have been also illustrated in human tumors, basically through activating mutations of the tyrosine kinase domain of these receptors: this is the scenario forFGFR1 in glioblastoma, for FGFR2 in endometrial cancers, for FGFR3 in bladder cancers and for FGFR4 in rhabdomyosarcoma. A t(8;9) (p11;q33) translocation of the FGFR1 gene is accountable for the spontaneous dimerization of the receptor in myeloproliferative syndromes. FGFR1 overexpression is a significant oncogenic event existent in the bulk of prostate cancer cells, and FGFR1 amplification is existent in non-small cell lung cancersand in hormone-resistant breast cancers.

Among the TKIs, few molecules specifically target FGFRs; for example, dovitinib also targets the VEGF receptors. Antibodies against FGFR3 and protein traps against FGFs are progressed for the treatment of cancers showing an oncogenic activation of the FGF–FGFR pathways.

1.3.4. Hepatocyte Growth Factor and the Mesenchymal-Epithelial Transition Receptor

Mesenchymal-epithelial transition ([MET] factor receptor tyrosine kinase) is a TKR that's consists of two chains, α and β, linked by a disulfide bridge, but it is synthesized as a distinctive polypeptide chain that is later cleaved by a proprotein convertase named furin. MET comprises a SEMA domain similar to those existent in semaphorins and four immunoglobulin-like domains. Hepatocyte growth factor (HGF) is also an αβ heterodimer resultant from a cleavage of a precursor by a serine proteinase, matriptase; it comprises two MET-binding domains.

MET activation is a common oncogenic event in numerous cancer sorts and plays a key role in metastasis initiation through epithelial-to-mesenchymal transition. This activation can be attained by HGF or MET gene overexpression, MET gene rearrangements and mutations, and production of autocrine activation of cancer cells.

Targeting the MET pathway can be done with monoclonal antibodies directed toward HGF or MET and with tyrosine kinase inhibitors with inconstant specificity toward MET; tivantinib has a better specificity against MET while crizotinib is active on the ALK receptor and cabozantinib is active on the VEGF and TIE receptors and.

1.3.5. Glial Cell Line-Derived Neurotrophic Factors and the Rearranged During Transfection Receptor

The rearranged during transfection (RET) receptor is complicated in the development of the neural crest. It is structurally linked to cadherins. It is stimulated by a protein complex comprising GFL (GDNF family ligand) and a coreceptor, GFRα (GDNF family receptors alpha). Germinal mutations of RET are linked to type II multiple endocrine neoplasia, and somatic mutations are found in medullary thyroid carcinomas. Numerous TKIs have exhibited activity against RET and vandetanib has been released in the market for the treatment of medullary thyroid cancer.

1.3.6. Anaplastic Lymphoma Kinase and Leukocyte Receptor Tyrosine Kinase Receptors

These receptors, anaplastic lymphoma kinase (ALK) and leukocyte receptor tyrosine kinase, were initially detected in lymphoma, especially ALK, whose gene is rearranged in anaplastic large T-cell lymphomas (t2;5

translocation). In addition, It is oncogenic in a small subgroup (around 5%) of non-small-cell lung tumors through gene rearrangement with echinoderm microtubule-associated protein-like 4. The ligands of these receptors are small proteoglycans linked to heparan sulfate, pleiotrophin (PTN) and midkin (MDK). Numerous efficient tyrosine kinase inhibitors against ALK have been developed and released: crizotinib and ceritinib are currently available for the management of the small cell lung tumors which possess ALK rearrangements.

1.3.7. Vascular Endothelial GFs and Their Receptors

Four VEGFs, called VEGFA to VEGFD, have been detected as well as the closely related placental growth factor, PGF or PlGF. They are released as homodimers, the monomers being linked by two disulfide bridges. There are three cognate TKRs, VEGFR1 (gene FLT1, FMS-like tyrosine kinase 1), VEGFR2 and VEGFR3 (gene FLT4, FMS-like tyrosine kinase 4). These GFs activate chiefly the proliferation and maturation of endothelial cells, which are normal, non-cancerous cells, adding to oncogenesis through the generation of the blood vessels required for tumor nutrition.

VEGFA detect and binds VEGFR1 and VEGFR2, VEGFB and PGF detect and bind only VEGFR1, while VEGFC and D detect and bind VEGFR3 (Figure. 8). Neuropilins 1 and 2 (NRP1 and 2) are coreceptors to VEGFA, capable of binding the growth factor without giving a signal. They are linked to the VEGFR1 and 2 receptors and show to be needed for their activity.

VEGFA is the chief regulator of arterial and venous angiogenesis, with VEGFR2 as the main mediator of its proangiogenic activity, while VEGFR3, activated by VEGFC and D, is the key for lymphangiogenesis. VEGFB and PGF are not indispensable to angiogenesis; they are supposed to augment the angiogenic signal given by VEGFA by hindering its binding site on VEGFR1, which has only little tyrosine kinase activity and cannot add to angiogenesis. As a result of the low expression of VEGFR1, homodimers have less possibility to take place than VEGFR1–VEGFR2 heterodimers, which have restricted action on endothelial cells' proliferation and have basically a role in angiogenesis process regulation. Additionally, a soluble form of VEGFR1, without tyrosine kinase activity, can sequester the VEGFs and exerts a role in angiogenesis downregulation.

Numerous molecular sorts of VEGFA have been recognized, resultant from alternative splicing. The chief one, VEGF 165, is deficient in the

Cancer, *4*, 361-370.
15. Hallberg, B., & Palmer receptor tyrosine kinas 2013;13:685–700.
16. Harper, S. J, & Bates angiogenic therapeu
17. Harris, R. C., Chun *Experimental Cell*
18. Heuckmann, J. M growth factor rec in lung cancer.
19. Holbro, T., Ci and their role 99-110.
20. Holbro, T., signaling r *and Toxic*
21. Hynes, N comple
22. Jorisse Burge of ac
23. Nev TC
24. Pe C
25.
26.

exon 6 product. The VEGFA 121 isoform, short in the domain encoded by exons 6 and 7, has a minute affinity to glycosaminoglycans and has, as a consequence, a much higher diffusibility in the extracellular space, while the VEGFA 186 and VEGFA 201 isoforms show high affinity for these compounds and keep at the surface of the cells. The affinity of these sorts for neuropilins also varies, VEGFA 121, for example, having no binding site for these coreceptors.

VEGF generation by tumor cells, which is activated by numerous signals such as hypoxia and inflammation, is accountable for the neoangiogenesis needed by tumors to grow further a few millimeters. The endothelial cell does not contribute itself to the cancer phenotype, and because of its genetic stability, the VEGFRs existing in its membrane do not show activating mutations as other tumor cell receptors such as EGFR exert.

An anti-VEGFA monoclonal antibody, bevacizumab, has been released, and its antiangiogenic activity is being utilized in the treatment of colorectal cancers and other malignancies. Another way to target this growth factor uses a pseudoantibody, VEGF trap or aflibercept, whose Fab portion has been substituted by the sequences of VEGFR1 and VEGFR2 complicated in binding to VEGFA, VEGFB and PGF; this leads to a high affinity of this trap to all three GFs. Antibodies against VEGFR2 are also in focus and ramucirumab has been recently gained approval for the treatment of gastric cancers.

Several tyrosine kinase inhibitors have been progressed, with dissimilar specificity for the VEGFRs and cross-inhibition of other receptors, especially of the PDGFR family. It is not established whether the group targeting displayed by these TKIs is an advantage or a disadvantage for cancer treatment. Sunitinib and sorafenib, which are approved chiefly for the treatment of kidney cancer and hepatocarcinoma, have specifically low specificity to VEGFRs. Novel compounds show better specificity and can better be figured as antiangiogenic drugs: this is the case of axitinib, vatalanib, and pazopanib and regorafenib, which exhibit cross-reactivity, with the KIT and PDGFR receptors.

1.3.8. Angiopoietins and the TIE Receptors

Angiopoietins are, as VEGFs, endothelial GFs and exert a role in angiogenesis, demonstrating their potential consideration in oncology. As for VEGF genes, angiopoietin gene transcription is stimulated, among other factors, by hypoxia, through the activation of the HIF transcription

FURTHER READI

1. Appleman, L. J. (20
cancer therapy. *Jou*
2. Beenken, A., &
pathophysiology
253.
3. Blume-Jensen,
Nature. *411*, 3
4. Blumensche
(2012). Tar
therapy. *Jo*
5. Bose, R.,
perspecti
Researc
6. Brooks
fibrob
canc
7. Bu
P.,
Y
a
8.
9.
1

2
CHAPTER

MAP KINASE PATHWAY

2.1 FROM RECEPTOR STIMULATION TO RAS ACTIVATION

The growth factor receptor-bound protein 2 (GRB2) is an adapter protein with SH2 and SH3 domains. Its SH2 domain detects specific phosphotyrosine residues of activated TKRs. This permits the employment at the membrane of a GDP–GTP exchange factor (GEF) named son of sevenless homologue 1 (SOS1), which substitutes GDP by GTP as a ligand of the membrane protein RAS. The RAS proteins represent the large protein family of small G-proteins, and GDP–GTP exchange on a small G-protein is usually mediated by GEFs. While the pathway EGFR–GRB2–SOS1–RAS is the best demonstrated, there are numerous other pathway downstream activations of tyrosine kinase receptors: other SH2 domain-containing adapter proteins are capable of binding phosphotyrosine residues of tyrosine kinase receptors. This is particularly the case for FGF receptors, for which a protein named FRS (FGF receptor substrate) is phosphorylated by the receptor and then detected, at the level of its phosphotyrosine residues, by GRB2 or another adapter protein.

Three RAS protein homologues exert comparable roles in numerous tissues: Kirsten RAS (KRAS RAS), Harvey RAS (HRAS) and neuroblastoma RAS (NRAS). All small G-proteins function equally: they show high affinity for GDP, which is substituted by GTP via a GEF. This exchange stimulates a change of conformation, triggering downstream activity (for example, RAS activation by GTP leads particularly to its binding to a serine/threonine kinase called RAF). Small G-proteins possess an intrinsic GTPase activity, permitting the retroconversion of GTP to GDP by the elimination of the GTP gamma phosphate group. This GTPase activity may be triggered by a GTPase-activating protein (GAP), which represents, therefore, the third member of the functional triad GEF–small G-protein–GAP, which is implicated in many cellular events outside the control of proliferation demonstrated.

There are other potentials for RAS activation. In the pathway opened by TKR activation, the GEF is SOS1 and the GAP is RASGAP (p120 RASGAP). Another GEF for RAS is RAS guanyl nucleotide-releasing protein (RASGRP), which can be stimulated by other molecules: inositol trisphosphate (IP3) through Ca^{2+} mobilization and diacylglycerol (DAG). These two molecules can be processed from phosphatidylinositol by phospholipase C gamma, which shows an SH2 domain permitting its activation by a TKR. IP3 and DAG are 'second messengers' that can also be formed by phospholipase C beta, which is stimulated via another signaling pathway involving receptors with seven transmembrane domains named large heterotrimeric G-protein-coupled receptors (GPCR).

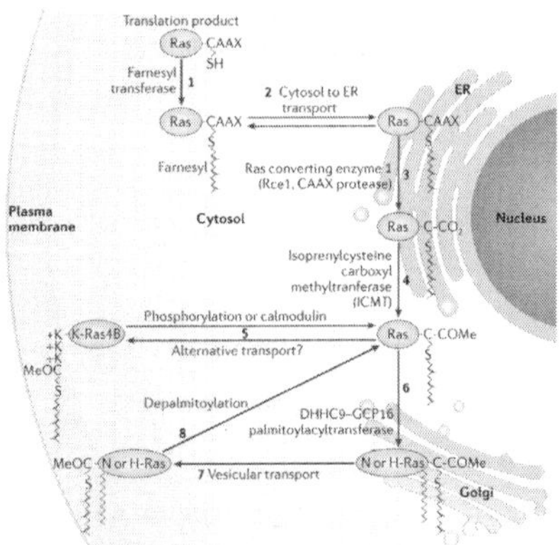

Figure. 10 Post-translational modifications of RAS proteins (Ahearn, Haigis, Bar-Sagi & Philips, 2012).

The RAS proteins are anchored to the plasma membrane through covalent bonds with a branched hydrophobic moiety established of 15 carbon atoms (farnesyl) or 20 carbon atoms (geranylgeranyl). These prenyl moieties, which are intermediates of cholesterol biosynthesis, are added as post-translational modifications at the level of the endoplasmic reticulum.

Their binding to the RAS proteins takes place on a cysteine residue of the polypeptidic chain, borne by a –CAAX motif (C for cysteine, A for aliphatic amino acid [Val, Leu or Ile], X for Met or Ser), the three amino acids AAX being removed after prenylation by RAS-converting endopeptidase (Figure. 10). In addition, some RAS proteins undergo palmitoylation in the Golgi apparatus. This modification is a covalent binding of palmitic acid, a fatty chain of 16 carbon atoms, at the level of a cysteine residue situated upstream the one that is prenylated. It is usually admitted that the enrollment of SOS1 to the membrane, owing to its affinity for GRB2, allows its interaction with its RAS protein substrate and, consequently, RAS activation. This membrane anchoring is needed for RAS activity.

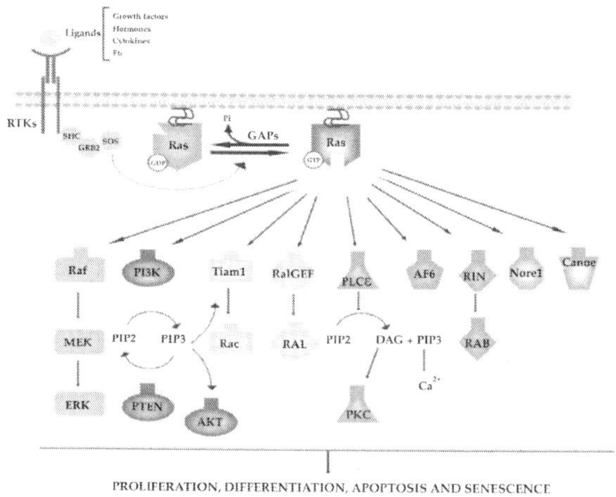

Figure. 11 Effectors of the RAS proteins (Castellano and Downward, 2010).

When RAS becomes bound to GTP, it undergoes a conformational modification that allows the recruitment to the membrane of a serine/threonine kinase which is an RAF protein in the MAP kinase pathway. Besides being the main activator of the MAP kinase pathway described below, RAS has many effector proteins (Figure. 11): RAS can directly trigger the catalytic subunit of PI3 kinase, in addition to other enzymes such as phospholipase C epsilon. It can also control the activity status of other small G-proteins by activating their GEFs or GAPs: this is the case for RAL, which exerts a role in endocytosis, and for RHO family G-proteins, which are implicated in the control of cytoskeleton movements. The RAS proteins are, therefore, at a significant hinge of signaling pathways. RAS protein activation is a quickly reversible, as intrinsic RAS GTPase activity, triggered by the RASGAP protein, rapidly deactivates RAS by reverting it back to its inactive form, RAS–GDP, which is incapable of recognizing RAF and attract it to the plasma membrane. RASGAP shows an SH2 domain permitting it to be recruited at the membrane level by a TKR to perform its action.

2.2 KINASE CASCADE

The direct target of RAS in the MAP kinase pathway is a serine/threonine kinase named RAF. There are three dissimilar RAF proteins, ARAF, BRAF and CRAF. RAF binding to RAS–GTP stimulates a conformational change which uncovers an activation loop permitting their phosphorylation and subsequently their activation. The activation process of the RAF proteins is complex and not fully decrypted; it implicates RAF dimerization and phosphorylation on numerous sites by numerous kinases such as cyclic AMP-activated protein kinase A, which is brought into exert following GPCR activation; the cytoplasmic tyrosine kinase SRC, triggered by numerous types of stimulus; the JAK kinase activated by cytokine receptors or the kinase called p21-activated kinase 1, which works after activation of the integrin or semaphorin pathways. RAF proteins may endure phosphorylation on numerous discrete sites. Two of them are in the N-terminal domain, on an SSYY motif where the first serine residue and the second tyrosine residue must be phosphorylated; nevertheless, for BRAF, the first serine residue is normally phosphorylated, and the two tyrosine residues are substituted by aspartic acid residues whose negative charge is constitutively 'phosphomimetic'. Two other phosphorylation sites are situated on a threonine and a serine residue (Thr 599 and Ser 602 for BRAF). While ARAF and CRAF need four phosphorylated residues for activity, BRAF needs only two; the consequences upon RAF-mediated oncogenesis will be completely below.

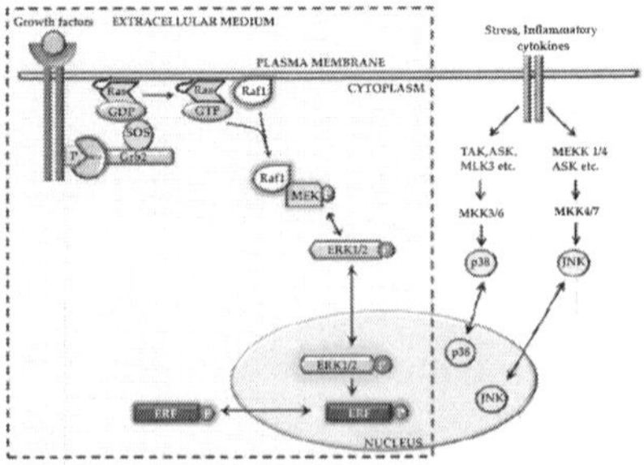

Figure. 12 MAP kinase pathways (Asai, 2002)

Two other phosphorylation sites are situated on a threonine and a serine residue (Thr 599 and Ser 602 for BRAF). While ARAF and CRAF need four phosphorylated residues for activity, BRAF needs only two; the significances upon RAF-mediated oncogenesis will be shown below. Downstream RAF kinases are the kinases MEK1 and MEK2, and downstream MEK1/2 are the kinases ERK1 and ERK2 (extracellular signal-regulated kinases). This represents, thus, a three-step kinase system, the RAF–MEK–ERK cascade. The RAF kinases are also named MAP3 kinases (MAP3K or MAPKKK), the MEK kinases are MAP2K or MAPKK and the ERK kinases are the suitable MAP kinases (MAPK). It will be revealed later that this RAF–MEK–ERK 'module' is only one among others and that other parallel MAP kinase pathways exist, with comparable three levels of phosphorylation. They are acting in stress response or inflammation relatively than in cell proliferation. A total of 25 MAP3K, 7 MAP2K and 13 MAPK have been identified in the superfamily of serine/threonine kinases.

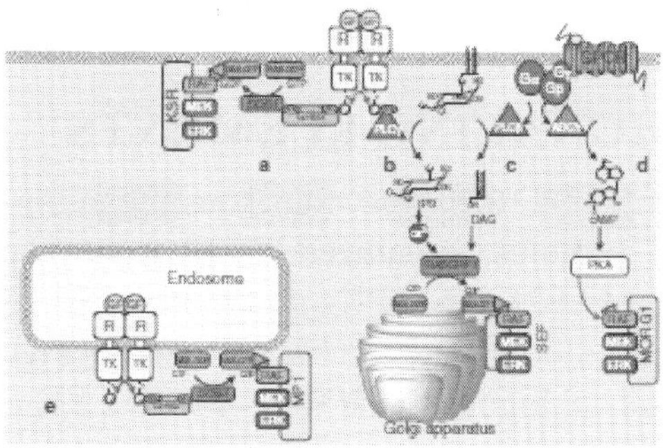

Figure. 13 Kinases organization and scaffold proteins (Robert, 2010).

MEK1/2 activation is attained by phosphorylation of two close serine residues: Ser 218 and Ser 222 for MEK1 and Ser 222 and Ser 226 for MEK2, situated in the activation loop of the catalytic center. The most active kinase on MEK1/2 is BRAF, tailed by CRAF, while ARAF has lower activity. MEK1/2 is in turn capable of phosphorylating ERK1/2. MEKs are dual kinases, capable of phosphorylating both a threonine and a tyrosine residue, also at the level of the activation loop, succeeding each other as TEY (Thr 202 –Glu 203 –Tyr 204 for ERK1 and Thr 185 –Glu 186 –Tyr 187 for ERK2). Kinase specificity to the kinase of lower level influenced by interactions between protein–protein binding domains rather than on different catalytic activities.

2.4 MAP KINASE SUBSTRATES

MAP kinases have the basal characteristic of phosphorylating various transcription factors, such as ELK1, ETS, FOS, JUN, MYC and SP1 which stimulate the expression of numerous genes implicated in cell cycle and cell proliferation. MAP kinases can also phosphorylate a cascade of kinases with numerous biological characters, known as MAPK-activated protein kinases, including ribosomal S6 kinases, mitogen and stress-activated kinases (MSKs) and MAP kinase-interacting proteins (MNKs). These kinases perform many effects in the cell, especially at the level of cell proliferation and adhesion.

2.4.1 MAP Kinases-Activated Transcription Factors MYC Cluster

The transcription factors of this group are characterized by a helix–loop–helix motif, which permits their dimerization with a common companion, MYC-associated factor X (MAX). Their activation is tightly regulated and their half-life is short; MAX is in comparison stable, ubiquitous and enduringly expressed. The proteins capable of binding MAX belong to numerous groups: the MYC group, originally recognized from the oncogenes c-myc, N-myc and L-myc and including the MYC, MYCN and MYCL proteins; the MAX dimerization protein group and the MNT (MAX-interacting protein) and MAX gene-associated proteins. The MYC–MAX complexes are transcriptional activators that bind promoter sequences comprising the hexanucleotide CACGTG of the E-box and employing the histone acetyltransferases needed for chromatin relaxation. They are also capable of repressing the transcription of other genes when binding the initiator elements (INR) of other gene promoters. Various genes are stimulated by MYC at the transcriptional level: genes implicated in cell growth and proliferation and in cell cycle. Among the genes whose transcription is inhibited by MYC are negative regulators of the cell cycle (p21 CIP1, p27 KIP1), as well as the gene encoding p53. In contradiction, MYC is capable of inducing apoptosis through BAX activation.

In distinction, the MAD–MAX complexes are transcriptional repressors at the level of the same gene promoters. This repression includes the corepressor SIN3 (silencing homologue 3), which belongs to the histone deacetylase complex. MAD activity stimulates an obstruction of cell proliferation and an inhibition of transformation, in addition to the induction of differentiation in various tissues.

FOS–JUN Axis

The transcription factors JUN and FOS (called after the corresponding oncogenes) belong to a family called activator protein 1 (AP1) and are linked to cell proliferation and differentiation. They operate after dimerization. There are three JUN proteins (JUN or c-JUN, JUNB and JUND) and four FOS proteins (FOS or c-FOS, FOSB, FRA1 and FRA2). The JUN proteins can produce homodimers or heterodimers with the FOS proteins, but the FOS proteins can only dimerize with JUN proteins. Heterodimerization with the transcription factors of the MAF (musculoaponeurotic fibrosarcoma) and activating transcription factor (ATF) clusters is also possible. The MAP kinases ERK, in addition to those of the JNK and p38 modules, are capable of phosphorylating the components of the AP1 complexes. Other phosphorylations are conceivable: for example, JUNB can be phosphorylated by the mitosis entry cyclin B–CDK1 complex which energies it to proteolysis.

The numerous dimers that can be generated may have contradictory physiological roles: the JUN–FOS dimer stimulates the transcription of proliferation genes, such as the genes of cyclin D (CCND) or of CDK4 and CDK6, whereas the JUN–JUNB or JUN–JUND dimers are incapable to do so; they seem, therefore, as MAP inhibitors. They stimulate, in distinction, the transcription of negative regulators of cell cycle. This demonstrates why the different MAP kinase modules, which differentially activate the several components of the AP1 complexes, may have contradictory effects on cell proliferation and differentiation. Among the target genes of the AP1 complexes are also those encoding cytokines and TNF-related factors.

ETS Family

This family includes a cascade of transcription factors, among which are ETS1, ETS2, ELK1, SAP1, SAP2, ERF, NET and ELK3. They can gather with other transcription factors to produce functional dimers. A first group (ETS1 and 2) can gather with AP1 family elements; the group including ELK1 and SAP1 and 2 (collectively called T-cell factor) is gathered with transcription factors of the SRF family (serum responsive element [SRE]-binding transcription factor) to activate SRE sites on DNA. A third group is entirely accountable for transcription repression. The ETS family factors target genes encoding other transcription factors, such as MYC and FOS, cell cycle regulatory proteins, antiapoptotic proteins, cytokines, growth factors, and so forth.

accessible for kinase inhibition do not appear to present direct anticancer effects. Nevertheless, as chronic inflammation is a factor-stimulating tumor progression, these compounds may interfere with oncogenesis, and they have been tried in oncology. Additionally, these compounds may interfere with the proangiogenic characteristics of the JNK and p38 pathways and, thus, merit attention. Lastly, even if the transcription factors, which characterize the final part of the MAP kinase pathway, establish potent bona fide targets, the progression of inhibitors is still in infancy. Peptidomimetics targeting the protein–protein interactions at the level of MYC–MAX or JUN–FOS, for example, might show therapeutic attention in oncology.

FURTHER READING

1. Ahearn, I. M., Haigis, K., Bar-Sagi, D., & Philips, M. R. (2012). Regulating the regulator: post-translational modification of RAS. *Nature Review Molecular Cell Biology*, *13*, 39-51.
2. Asai, T., Tena, G., Plotnikova, J., Willmann, M. R., Chiu, W. L., Gomez-Gomez, L., Boller, T., Ausubel, F. M., & Sheen, J. (2002). MAP kinase signalling cascade in Arabidopsis innate immunity. *Nature*, *415*(6875), 977-983.
3. Berndt, N., Hamilton, D., & Sebti, S. M. (2011). Targeting protein prenylation for cancer therapy. *Nature Review Cancer*, *11*, 775-791.
4. Castellano, E., & Downward, J. (2010). Role of RAS in the regulation of PI 3-kinase. *Current Topics in Microbiology and Immunology*. *346*, 143-169.
5. Chang, F., Steelman, L. S., Lee, J. T, Shelton, J. G., Navolanic, P. M., Blalock, W. L., Franklin, R. A., & McCubrey, J. A. (2003) Signal transduction mediated by the Ras/Raf/MEK/ERK pathway from cytokine receptors to transcription factors: potential targeting for therapeutic intervention. *Leukemia*, *17*, 1263-1293.
6. Dhanasekaran, D. N., Kashef, K., Lee, C. M., Xu, H., & Reddy, E. P. (2007). Scaffold proteins of MAP-kinase modules. *Oncogene*, *26*, 3185-3202.
7. Dhillon, A. S., Hagan, S., Rath, O., & Kolch, W. (2007) MAP kinase signalling pathways in cancer. *Oncogene*, *26*, 3279-3290.
8. Karnoub, A. E., & Weinberg, R. A. (2008). Ras oncogenes: split personalities. *Nature Reviews Molecular Cell Biology*, *9*, 517-531.
9. Konstantinopoulos, P. A., Karamouzis, M. V., & Papavassiliou, A. G. (2007). Post-translational modifications and regulation of the RAS superfamily of GTPases as anticancer targets. *Nature Reviews Drug Discovery*, *6*, 541-555.
10. Leicht, D. T, Balan, V., Kaplun, A., Singh-Gupta, V., Kaplun, L., Dobson, M., & Tzivion, G. (2007). Raf kinases: function, regulation and role in human cancer. *Biochimica et Biophysica Acta*. *1773*, 1196-1212.
11. McCubrey, J. A., Steelman, L. S., Abrams, S. L, Bertrand, F. E., Ludwig, D. E., Bäsecke, J., Libra, M., Stivala, F., Milella, M., Tafuri, A., Lunghi, P., Bonati, A., & Martelli, A. M. (2008). Targeting survival cascades induced by activation of Ras/Raf/MEK/ERK, PI3K/PTEN/

Akt/mTOR and Jak/STAT pathways for effective leukemia therapy. *Leukemia*, *22*, 708-722.

12. McKay, M. M., & Morrison, D. K. (2007). Integrating signals from RTKs to ERK/MAPK. *Oncogene*, *26*, 3113-3121.
13. Michaloglou, C., Vredeveld, L. C., Mooi, W. J., & Peeper, D. S. (2008). BRAF (E600) in benign and malignant human tumours. *Oncogene*, *27*, 877-895.
14. Montagut, C., & Settleman, J. (2009). Targeting the RAF-MEK-ERK pathway in cancer therapy. *Cancer Letters*, *283*, 125-134.
15. Pritchard, A. L., & Hayward, N. K. (2013). Molecular pathways: mitogen-activated protein kinase pathway mutations and drug resistance. *Clinical Cancer Research*, *19*, 2301-2309.
16. Pylayeva-Gupta, Y., Grabocka, E., & Bar-Sagi, D. (2011). RAS oncogenes: weaving a tumorigenic web. *Nature Reviews Cancer*, *11*, 761-774.
17. Robert, J. (2010). Textbook of cell signalling in cancer, an educational approach. Paris, France: Springer-Verlag.
18. Roberts, P. J, & Der, C. J. (2007). Targeting the Raf-MEK-ERK mitogen-activated protein kinase cascade for the treatment of cancer. *Oncogene*, *26*, 3291-3310.
19. Roux, P. P. (2004). Blenis J.ERK and p38 MAPK-activated protein kinases: a family of protein kinases with diverse biological functions. *Microbiology and Molecular Biology Reviews*, *68*, 320-344.
20. Santarpia, L., Lippman, S. M., & El-Naggar, A. K. (2012). Targeting the MAPK-RAS-RAF signaling pathway in cancer therapy. *Expert Opinion on Therapeutic Targets*, *16*, 103-119.
21. Sebolt-Leopold, J. S., Herrera, R. (2004). Targeting the mitogen-activated protein kinase cascade to treat cancer. *Nature Reviews Cancer*, *4*, 937-947.
22. Steelman, L. S., Abrams, S. L., Whelan, J., Bertrand, F. E., Ludwig, D. E., Bäsecke, J., Libra, M., Stivala, F., Milella, M., Tafuri, A., Lunghi, P., Bonati, A., Martelli, A. M., & McCubrey, J. A. (2008). Contributions of the Raf/MEK/ERK, PI3K/PTEN/Akt/mTOR and Jak/STAT pathways to leukemia. *Leukemia*, *22*, 686-707.

3
CHAPTER

PHOSPHATIDYLINOSITOL 3-KINASE PATHWAY

3.1.FROMPHOSPHATIDYLINOSITOL3-KINASETO AKT PROTEINS

Phosphatidylinositol 3-kinases (PI3Ks) are a group of lipid kinases linked to serine/threonine kinases. They work as heterodimers comprising a catalytic subunit and a regulatory subunit. In the main PI3Ks, the catalytic subunit (p110) is encrypted by the PIK3CA gene, and the regulatory subunit (p85) is encoded by the PIK3R1 gene. This regulatory subunit possesses an SH2 domain permitting binding to phosphotyrosine residues of activated TKRs, which trigger the activation of the catalytic subunit (Figure. 15). In some cases, particularly for the growth factors of the insulin family, p85 activation takes place via an adapter protein, IRS1 or 2 (insulin receptor substrate 1 or 2); this protein is phosphorylated by the IGF1R receptor after binding to phosphotyrosine residues, and the phosphotyrosine residues, therefore, produced on IRSs are recognized by the SH2 domain of p85.

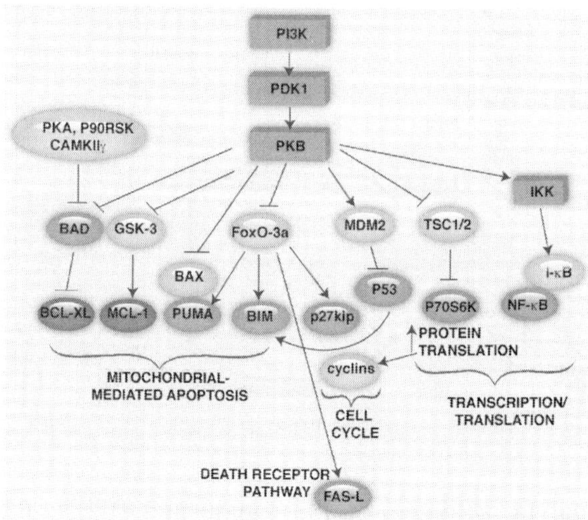

Figure. 17 PI3 kinase pathway after activation by a TKR (Duronio, 2008).

Among the chief substrates of AKT proteins, and without the probability of being exhaustive, one can remark (Figure. 17):

- The tuberous sclerosis complex 2 (TSC2) protein, also called tuberin. This protein is inhibited by AKT-induced phosphorylation on Ser 939 and Thr 1462 residues; once combined with TSC1 (hamartin), the complex represents a GTPase-activating protein for a small G-protein of the RAS family named RAS homologue enriched in the brain (RHEB); this pathway result in mechanistic target of rapamycin (mTOR) activation.
- BCL2-associated agonist of cell death (BAD), which is a 'BH3-only' protein implicated in the positive regulation of mitochondrial apoptosis; this protein is inactivated by phosphorylation.
- MDM2 (murine double-minute p53 binding) protein, stimulated by phosphorylation, which is the E3 ubiquitin ligase of protein p53; MDM2 activation is therefore in contradiction to the p53 effects on cell cycle arrest and apoptosis initiation.
- p21 CIP1 and p27 KIP1, which are cyclin-dependent kinase (CDK) inhibitors and therefore are negative regulators of the progression of cells in the G1 phase of the cycle; AKT-mediated phosphorylation hinders these CDK inhibitors.
- Glycogen synthase kinase 3β (GSK3β), which incorporates signals of metabolic origin and which exerts a significant role

in the Wnt pathway. This protein is repressed by AKT-mediated phosphorylation; as a result, GSK3β can no longer phosphorylate β-catenin, and this last protein can relocalize in the nucleus and activate transcription programs resulting in cell proliferation and cell cycle entry, as a substitute of being directed to the proteasome for proteolysis.

- IKK (IκB kinase), stimulated by phosphorylation, which phosphorylates and constrains the IκB (inhibitor of NFκB) protein, an inhibitor of the transcription factor NFκB, as a result, AKT positively regulates NFκB, which is implicated in cell survival, proliferation, invasion, angiogenesis and chemoresistance.
- WEE1, repressed by this phosphorylation, which is a CDK1-inhibiting kinase, a chief component of mitosis triggering.
- Apoptosis signal-regulating kinase 1, repressed by AKT-mediated phosphorylation, which is a MAP3K activating the JNK pathway implicated in apoptosis induction in reaction to proapoptotic signals and to endothelial reticulum stress.
- Forkhead box class O (FOXO), which are apoptosis-inducing transcription factors working on both the extrinsic pathway (TRAIL, FASL) and on the mitochondrial pathway (BH3-only proteins); their phosphorylation by AKT on Thr 32, Ser 253 and Ser 315 stimulates their cytoplasmic relocalization and the loss of their transcriptional functions.

All AKT actions, consequently, trigger the cells toward proliferation and survival.

3.3. mTOR PROTEIN AND THE TORC COMPLEXES

The mTOR protein was called by analogy with a yeast protein that is repressed by a natural product, rapamycin, after binding to a small protein, FKBP12 (FK506-binding protein 12). This is a serine/threonine kinase working downstream AKT and is implicated in the phosphorylation of a diversity of substrates. mTOR, in addition to the PI3Ks, belongs to the cluster of PIKKs (phosphoinositide 3-kinase-related protein kinases), like ATM and ATR implicated in DNA repair. A scheme of the numerous domains of the protein is shown in Figure. 3.4. mTOR is not directly activated by AKT: its activation follows the action of the small G-protein RHEB, which is active when combined with GTP. The double inhibition of TSC2 by AKT and of

FURTHER READING

1. Abraham, R. T., & Gibbons, J. J. (2007). The mammalian target of rapamycin signaling pathway: twists and turns in the road to cancer therapy. *Clinical Cancer Research*, *13*, 3109-3114.
2. Bader, A.G., Kang, S., Zhao, L., & Vogt, P. K. (2005). Oncogenic PI3K deregulates transcription and translation. *Nature Reviews Cancer*, *5*, 921-929.
3. Bartholomeusz, C., & Gonzalez-Angulo, A. M. (2012) Targeting the PI3K signaling pathway in cancer therapy. *Expert Opinion on Therapeutic Targets*, *16*, 121-130.
4. Burris, 3rd H. A. (2013). Overcoming acquired resistance to anticancer therapy: focus on the PI3K/AKT/mTOR pathway. *Cancer Chemotherapy and Pharmacology*, *71*, 829-842.
5. Chalhoub, N., & Baker, S. J. (2009). PTEN and the PI3-kinase pathway in cancer. *Annual Review of Pathology*, *4*, 127-150.
6. Coquillard, C., Valery, V., Marti, F., & Gedaly, R. (2015). mTOR signaling in regulatory T cell differentiation and expansion. *SOJ Immunology*, *3*(1), 1-10.
7. Courtney, K. D., Corcoran, R. B., & Engelman, J. A. (2010). The PI3K pathway as drug target in human cancer. *Journal of Clinical Oncology*, *28*, 1075-1083.
8. Cully, M., You, H., Levine, A. J., & Mak, T. W. (2006). Beyond PTEN mutations: the PI3K pathway as an integrator of multiple inputs during tumorigenesis. *Nature Review Cancer*, *6*, 184-192.
9. Dancey, J. (2010). MTOR signaling and drug development in cancer. *Nature Reviews Clinical Oncology*, *7*, 209.
10. Duronio, V. (2008). The life of a cell: apoptosis regulation by the PI3K/PKB pathway. *Biochemical Journal*, *415*(3), 333-344.
11. Engelman, J. A., Luo, J., & Cantley, L. C. (2006). The evolution of phosphatidylinositol 3-kinases as regulators of growth and metabolism. *Nature Reviews Genetics*, *7*, 606-619.
12. Fu, Z., & Tindall, D. J. (2008). FOXOs, cancer and regulation of apoptosis. *Oncogene*, 27, 2312-2319.
13. Grabiner, B. C., Nardi, V., Birsoy, K., Possemato, R., Shen, K., Sinha, S., Jordan, A., Beck, A. H., & Sabatini, D. M. (2014). A diverse array of cancer-associated mTOR mutations are hyperactivating and can

predict rapamycin sensitivity. *Cancer Discovery*, *4*, 554-563.

14. Guertin, D. A., & Sabatini, D. M (2007). Defining the role of mTOR in cancer. *Cancer Cell*, *12*, 9-22.
15. Ihle, N. T., & Powis, G. (2009). Take your PIK: phosphatidylinositol 3-kinase inhibitors race through the clinic and toward cancer therapy. *Molecular Cancer Therapeutics*, *8*, 1-9.
16. Inoki, K., Corradetti, M. N., & Guan, K. L. (2005). Dysregulation of the TSC-mTOR pathway in human disease. *Nature Genetics*, *37*, 19-24.
17. Inoki, K., Kim, J., & Guan, K.L. (2012). AMPK and mTOR in cellular energy homeostasis and drug targets. *Annual Review Of Pharmacology And Toxicology*, *52*, 381-400.
18. Jewell, J. L., Russell, R. C., & Guan, K. L. (2013). Amino acid signalling upstream of mTOR. *Nature Reviews Molecular Cell Biology*, *14*, 133-139.
19. Keniry, M., & Parsons, R. (2008). The role of PTEN signaling perturbations in cancer and in targeted therapy. *Oncogene*, *27*, 5477-5485.
20. Laplante, M., & Sabatini, D. M. (2013). Regulation of mTORC1 and its impact on gene expression at a glance. *Journal of Cell Science*, *126*, 1713-1719.
21. Liu, P., Cheng, H., Roberts, T. M., & Zhao, J. J. (2009). Targeting the phosphoinositide 3-kinase pathway in cancer. *Nature Reviews Drug Discovery*, *8*, 627-644.
22. Meng, D., Frank, A. R., & Jewell, J. L. (2018). mTOR signaling in stem and progenitor cells. *Development*, *145*.
23. Meric-Bernstam, F., & Gonzalez-Angulo, A. M. (2009). Targeting the mTOR signaling network for cancer therapy. *Journal of Clinical Oncology*, *27*, 2278-2287.
24. Perl, A. (2016). Activation of mTOR (mechanistic target of rapamycin) in rheumatic diseases. *Nature Reviews Rheumatology*, *12*, 169-182.
25. Robert J. (2010). Textbook of cell signalling in cancer, an educational approach. Paris, France: Springer-Verlag.
26. Rodon, J., Dienstmann, R., Serra, V., & Tabernero, J. (2013). Development of PI3K inhibitors: lessons learned from early clinical trials. *Nature Reviews Clinical Oncology*, *10*, 143-153.

27. Sengupta, S., Peterson, T. R., & Sabatini, D. M. (2010). Regulation of the mTOR complex 1 pathway by nutrients, growth factors, and stress. *Molecular Cell*, *40*, 310-322.
28. Sparks, C. A., & Guertin, D. A. (2010).Targeting mTOR: prospects for mTOR complex 2 inhibitors in cancer therapy. *Oncogene*, *29*, 3733-3744.
29. Vanhaesebroeck, B., Stephens, L., & Hawkins, P. (2012). PI3K signalling: the path to discovery and understanding. *Nature Reviews Molecular Cell Biology*, *13*, 195-203.
30. Yang, J. Y., & Hung, M. C. (2009). A new fork for clinical application: targeting forkhead transcription factors incancer. *Clinical Cancer Research*, *15*, 752-757.
31. Yap, T. A., Garrett, M. D., Walton, M. I., Raynaud, F., de Bono, J. S, & Workman, P. (2008). Targeting the PI3K-AKTmTOR pathway: progress, pitfalls, and promises. *Current Opinion in Pharmacology*, 8, 393-412.
32. Yuan, T. L., Cantley, L. C. (2008). PI3K pathway alterations in cancer: variations on a theme. *Oncogene*, *27*, 5497-5510.

4
CHAPTER

G-PROTEIN-COUPLED RECEPTORS

4.1. STRUCTURE AND MECHANISM OF ACTION OF G-PROTEIN-COUPLED RECEPTORS

4.1.1. Receptors

G protein-coupled receptors (GPCRs) comprised a single polypeptidic chain equipped with seven transmembrane domains; these domains are made of a cascade of 22–25 hydrophobic amino acids, which permit the formation of an α-helix inserted within the aliphatic chains on membrane phospholipids. The N-terminal part of the receptors is extracellular while the C-terminal part is intracellular; three loops separating the transmembrane domains existing on each side of the membrane. The count of receptors in this family is higher than 800, which represents an important part of the genome; they are divided in numerous families and subfamilies as a function of their structure, their ligands and the type of binding, as seen below:

- The rhodopsin receptors family, Among them, are receptors to small molecules with intercellular messenger function: acetylcholine, catecholamines and other biogenic amines; nucleosides and nucleotides; lipid mediators such as retinal, prostaglandins, leukotrienes and thromboxanes, free fatty acids and endocannabinoids; and protein and peptide hormones.

- This family includes also all the olfactory and gustatory receptors, which represent more than 400 diverse proteins. The binding between receptor and ligand includes deep transmembrane structures of the receptors. Some of these receptors are stimulated by thrombin- or trypsin-induced proteolysis, which cleaves the N-terminal part of the receptor, which then behaves as an autonomous ligand; they are named protease-activated receptors.
  - The secretin receptors family (15 members), which includes the receptors to numerous peptide hormones: glucagon, parathormone, calcitonin and hypothalamic releasing factors (corticotropin-releasing hormone, gonadotropin-releasing hormone or luteinizing hormone-releasing hormone, thyrotropin-releasing hormone).
- The glutamate receptors family (22 members), including the metabotropic receptors (as opposed to ionotropic receptors, of several neurotransmitters, which bind to the N-terminal part of the receptor. This family also includes a Ca 2+ 'sensor' (CASR); the calcium ion binds to an extension of the N -terminal part of the receptor, which then interacts with the deep transmembrane domains and behaves as an 'auto-ligand'.
- The adhesion receptors family, whose receptors have a bulky N-terminal domain permitting them to interact with extracellular molecules, but whose true ligands are chiefly proteoglycans.
  - The frizzled receptors family, which includes the 10 WNT protein receptors and the second receptor of the Hedgehog pathway, named smoothened (SMO). These receptors show structural analogies with GPCR, but the G-protein activation distinguishing for GPCR has not been always illustrated.

4.1.2. G-Proteins

The collective final way for GPCR activation resides in the implication of large heterotrimeric G-proteins. All G-proteins are capable of binding and hydrolyzing GTP; there is a distinction between 'small G-proteins' such as RAS, RHEB or RAG and 'large heterotrimeric G-proteins' which are transducers of the information received by GPCR. Large heterotrimeric G-proteins contain three polypeptidic chains, called α, β and γ. The β and γ chains are tightly bound in a special entity βγ. G-proteins are anchored to the membrane by two lipid moieties, the first one being a linear chain, which is added post-translationally at the N-terminal extremity of the α subunit, and the second one being a branched chain, which is post-translationally added to the C-terminal part of the γ subunit. The α subunit contains the binding domain of guanyl nucleotides, which is busy by GDP at basal state. Large heterotrimeric G-proteins are stably linked to an effector enzyme and can transiently associate with a GPCR.

GPCR ligand binding stimulates this transient association between a GPCR and a large G-protein, and this binding stimulates the replacement of GDP by GTP at the level of the α subunit. The α subunit then detaches from the βγ subunit and realizes the GTP cleavage owing to its GTPase activity. During the brief period when the α subunit is bound to GTP, the linked effector enzyme is stimulated and can exert its catalytic activity. As soon as GTP is hydrolyzed, this regulation stops, the effector enzyme returns to basal state and the α subunit relinks to the βγ subunit.

Likewise, the βγ subunit can stimulate other effectors before relinking with the GDP-bound α subunit. Large G-proteins, therefore, work as switches yielding one signal only; they must relink to a newly stimulated receptor molecule and exchange GDP against GTP to give a new impulse to their effectors.

In front of hundreds of receptors, there is a count of 16 genes encoding the α subunits, 5 for β subunits and 12 for the γ subunits of G-proteins. The α subunits mirror the tissue and functional variety of G-proteins, while the βγ subunits are rather implicated in controlling G-protein activity. There are numerous sorts of α subunits, the main ones being α s, α i, α q and α 12, which differ accordingly to the effector enzyme they can stimulate: GTP-bound α s subunits activate adenylyl cyclases (ADCYs), while α subunits impede this activity; GTP-bound α q subunits stimulate phospholipases C (PLCs) beta and GTP-bound α 12 subunits stimulate small G-proteins of the RHO family. Adenylyl cyclase and PLCβ produce second messengers, which are shared intermediates of the signals brought by GPCR ligands. The βγ subunit, momentarily detached from the α subunit during activation, also has suitable effectors, particularly the regulatory subunit of PI3 kinase gamma, as well as some sorts of ionic channels.

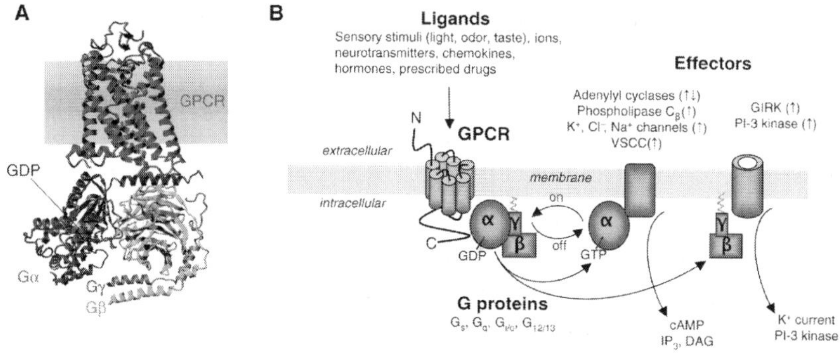

Figure. 21 Activation and function of large heterotrimeric G-proteins (Vilarda-ga Agnati, Fuxe & Ciruela, 2010).

The action of large heterotrimeric G-proteins is regulated by RGS (regulator of G-protein signaling) proteins, which activate GTPase activity and are thus GAPs (GTPase-activating proteins). Consequently, they deactivate large G-proteins that are brought back to GDP-bound, inactive state. There are about 20 RGS proteins with adjustable specificity toward G-proteins. Large heterotrimeric G-proteins can also be regulated by phosphorylation: the βγ subunit is capable of recruiting a kinase which phosphorylates the stimulated receptor and accelerates its desensitization via the formation of a β-arrestin-binding site, inducing receptor internalization and recycling.

4.2. SECOND MESSENGERS OF GPCR ACTIVATION

4.2.1. Adenylyl Cyclases and Cyclic AMP

ADCYs are high-molecular-weight membrane enzymes, comprising two homologous parts, each made of six transmembrane helices and a large C-terminal domain named the C-domain. The two C-domains are linked and show an ATP-binding site and a catalytic site for its conversion into cAMP. These enzymes are regulated by heterotrimeric large G-proteins; those equipped with an α s subunit activate adenylyl cyclase when bound to GTP, while those fortified with an α subunit inhibit adenylyl cyclase. They are also regulated by numerous factors and integrate much information, particularly through Ca^{2+} binding and protein kinase C-induced phosphorylation.

Cyclic APM (cAMP) is the second messenger of a large count of signals established by GPCR: hormones, neurotransmitters and so forth. It is implicated in the regulation of numerous enzymes, particularly in metabolic pathways. cAMP targets are protein kinases A, which are tetrameric serine/threonine kinases comprising two catalytic and two regulatory subunits. They phosphorylate their targets on the serine residues of an RRXS consensus motif. cAMP-induced PKA activation is cooperative, which permits a rapid and widespread effect, rapidly terminated by the fact that PKAs phosphorylate and activate cAMP phosphodiesterases, which hydrolyze cAMP. PKAs target various proteins, among which are glycogen phosphorylase kinase, which is stimulated, and glycogen synthase, which is deactivated, in reaction to this phosphorylation. PKAs also play an effect at the gene transcription level, by activating through phosphorylation a transcription factor named cAMP response element-binding protein, which relocalizes into the nucleus and detects specific promoter sequences of numerous target genes, named cAMP response element. cAMP is also capable of directly activating exchange protein directly activated by cAMP (EPAC), a GTP–GDP exchange factor of a small G-protein, RAP, which stimulates other intracellular signaling pathways.

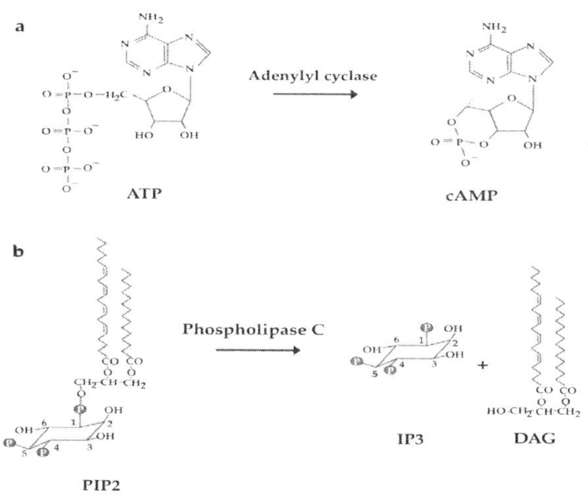

Figure. 22 Formation of second messengers (Robert, 2010).

4.2.2. Phospholipases C, Diacylglycerol and Inositol 1,4,5-Tris-phosphate

PLCs are membrane enzymes armed with a pleckstrin homology domain permitting them to recognize and bind phosphoinositides. PLCγ can be activated by a tyrosine kinase receptor, while PLCβs are stimulated by a

neurology and oncology. PKCs are involved in proliferation and PKCβ and PKCδ in angiogenesis; PKCε is overexpressed in breast and colon cancers, PKCη in non-small cell lung cancers and glioblastomas.

FURTHER READING

1. Audet, M., & Bouvier, M. (2012). Restructuring G-protein–coupled receptor activation. *Cell, 151*, 14-23.
2. Balakin, K. V., Ivanenkov, Y. A., Tkachenko, S. E., Kiselyov, A. S., & Ivachtchenko, A. V. (2008). Regulators of chemokine receptor activity as promising anticancer therapeutics. *Current Cancer Drug Targets*, *8*, 299-340.
3. Berridge, M. J. (2009). Inositol trisphosphate and calcium signalling mechanisms. *Biochimica et Biophysica Acta*, *1793*, 933-940.
4. Deng, L., Chen, N., Li, Y., Zheng, H., & Lei, Q. (2010). CXCR6/ CXCL16 functions as a regulator in metastasis and progression of cancer. *Biochimica et Biophysica Acta*, *1806*, 42-49.
5. Dhanasekaran, N., Tsim, S. T., Dermott, J. M., & Onesime, D. (1988). Regulation of cell proliferation by G proteins. *Oncogene*, *17*, 1383-1394.
6. Domanška, U. M., Kruizinga, R. C., Nagengast, W. B., Timmer-Bosscha, H., Huls, G., de Vries, E. G., & Walenkamp, A. M. (2013). A review on CXCR4/CXCL12 axis in oncology: no place to hide. *European Journal of Cancer*, *49*, 219-230.
7. Entschladen, F., Zänker, K. S., & Powe, D. G. (2011). Heterotrimeric G protein signaling in cancer cells with regard to metastasis formation. *Cell Cycle*, *10*, 1086-1091.
8. Fraser, C. C. G (2008). Protein-coupled receptor connectivity to NF-kappaB in inflammation and cancer. *International Reviews of Immunology*, *27*, 320-350.
9. Garin, A., & Proudfoot, A. E. (2011). Chemokines as targets for therapy. *Experimental Cell Research*, *317*, 602-612.
10. Heasley, L. E. (2011). Autocrine and paracrine signaling through neuropeptide receptors in human cancer. *Oncogene*, *20*, 1563-1569.
11. Houshmand, P., & Zlotnik, A. (2003). Therapeutic applications in the chemokine superfamily. *Current Opinion in Chemical Biology*, *7*, 457-460.
12. Karnoub, A. E., & Weinberg, R. A. (2006-2007). Chemokine networks and breast cancer metastasis. *Breast Disease*, *26*, 75-85.
13. Khan, S. M., Sleno, R., Gora, S., Zylbergold, P., Laverdure, J. P., Labbé, J. C., Miller, G.J., & Hébert, T.E. (2013). The expanding roles

of Gβγ subunits in G protein-coupled receptor signaling and drug action. *Pharmacological Reviews*, *65*, 545-577.

14. Lagerstrom, M. C., & Schioth, H. B. (2008). Structural diversity of G protein-coupled receptors and significance for drug discovery. *Nature Reviews Drug Discovery*, *7*, 339-357.
15. Lappano, R., Maggiolini, M. G. (2011). Protein-coupled receptors: novel targets for drug discovery in cancer. *Nature Reviews Drug Discovery*, *10*, 47-60.
16. Marinissen, M. J., & Gutkind, J. S. (2001). G-protein-coupled receptors and signaling networks: emerging paradigms. *Trends in Pharmacological Sciences*, *22*, 368-376.
17. O'Hayre, M., Vázquez-Prado, J., Kufareva, I., Stawiski, E. W., Handel, T. M., Seshagiri, S., & Gutkind, J. S. (2013). The emerging mutational landscape of G proteins and G-protein-coupled receptors in cancer. *Nature Reviews Cancer*, *13*, 412-424.
18. Olson, T. S., & Ley, K. (2002). Chemokines and chemokine receptors in leukocyte trafficking. *American Journal of Physiology*, *283*, R7-R28.
19. Pisanti S, Picardi P, D'Alessandro A, Laezza C, Bifulco M. The endocannabinoid signaling system in cancer. Trends Pharmacol Sci. 2013;34:273–82.
20. Robert, J. (2010). Textbook of cell signalling in cancer, an educational approach. Paris, France: Springer-Verlag.
21. Teicher, B. A., & Fricker, S. P. (2010). CXCL12 (SDF-1)/CXCR4 pathway in cancer. *Clinical Cancer Research*, *16*, 2927-2931.
22. Toker, A. (1998). Signaling through protein kinase C. *Frontiers in Bioscience*, *3*, d1134-d1147.
23. Vilardaga, J., Agnati, L. F., Fuxe, K., & Ciruela, F. (2010). G-protein-coupled receptor heteromer dynamics. *Journal of Cell Science*, *123*, 4215-4220.
24. Wang, Z., & Huang, H. (2013). Platelet factor-4 (CXCL4/PF-4): an angiostatic chemokine for cancer therapy. *Cancer Letter*, *331*, 147-153.
25. Waugh, D. J., & Wilson, C. (2008). The interleukin-8 pathway in cancer. *Clinical Cancer Research*, *14*, 6735-6741.
26. Zlotnik, A., Yoshie, O., & Nomiyama, H. (2006). The chemokine and chemokine receptor superfamilies and their molecular evolution. *Genome Biology*, *7*, 243.

5 CHAPTER

TGFβ PATHWAY

5.1 TGFβ FAMILY LIGANDS

These factors and their receptors may have very flexible tissue distribution and are expressed at certain periods of life: some of them are expressed only in some cell sorts and/or during very short times of embryonic life, while others are ubiquitous and/or expressed lifelong.

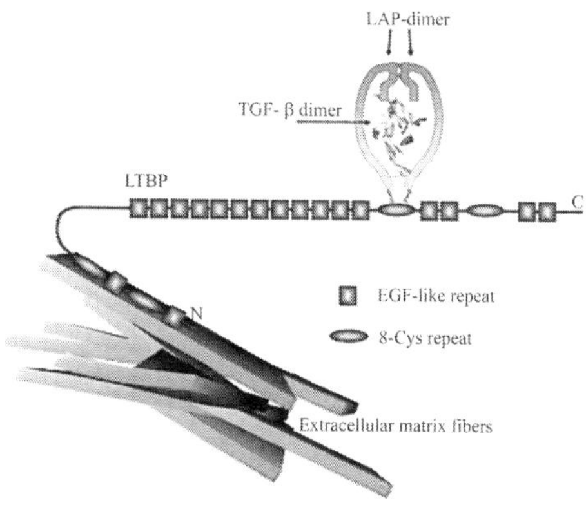

Figure 25 TGFβ dimer (Garcia-Fernandez and Molina, 2009).

In the TGFβ subfamily exist three TGFβs; a chain of proteins named inhibins or activins, nodal, lefty and myostatin; in addition to GDF (growth differentiation factors) 1 and 3. In the BMP subfamily exist BMP 2, 4, 5, 6, 7, 9 and 10; GDF 5, 6 and 7; and anti-Mullerian hormone (AMH). These factors are synthesized as precursor latent forms, which are subsequently submitted to proteolytic cleavage liberating the active C-terminal part of the protein. The proteolytic enzymes responsible for this activation are convertases of the subtilisin family, such as furin, and together named proprotein convertases subtilisin/kexin or paired basic amino acid cleaving enzymes. After cleavage, the three TGFβs stay non-covalently attached to the N-terminal peptide, named the latency-associated peptide, which is not the situation for other factors of the TGFβ and BMP subfamilies. All TGFβs and BMPs are released as dimers and are characterized by the existence of a distinguishing cysteine knot at the C-terminal level, which permits the formation of three intramolecular disulfide bridges; the last disulfide bridge is utilized for dimerization (Figure 25).

All these factors are characterized by an antiparallel dimeric structure, which is formed before receptor interaction, permitting the recruitment of two molecules of type II receptors and of two molecules of type I receptor (Fig. 26). The ligands of the TGFβ subfamily have more affinity for type II receptors, and this is only afterward that two molecules of type I receptor are engaged and bound to the ligand type II receptor complex. Some ligands of the BMP subfamily can bind indifferently to any receptor sort; the complex, therefore, formed engaging afterwards the receptor of the other type. Dimerization usually takes place between two identical molecules (homodimerization); however, the dimerization between an inhibin (INHA) and an activin (INHB) is probable, the heterodimer harboring a negative effect on receptor recruitment.

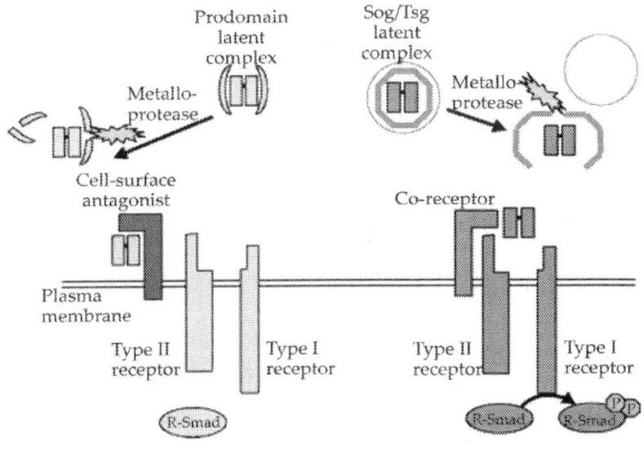

Figure. 26 TGFβ reception complex (Whitman and Raftery, 2005).

In the extracellular space, ligands of both subfamilies are protected from receptor interaction by antagonistic proteins, which are usually secreted proteins, each of them being more or less specific of the factors. Some of them are linked to glycans to form proteoglycans, such as decorin; others are structural analogues of the ligands and inhibit their interaction with their receptor through a mechanism of negative dominance (cerberus, gremlin and sclerostin); others are binding proteins whose progressive diffusion in the extracellular space produces gradients, which exert a chief role in morphogenetic processes: this is the situation of follistatin, chordin or noggin.

5.2. RECEPTORS ACTIVATION

At the level of the target cells of this signaling pathway, the receptors of the TGFβ and BMP subfamilies are membrane proteins with a sole transmembrane domain, an extracellular portion aimed at ligand detection and binding and an intracellular part harboring serine/threonine kinase activity. These receptors are of two types, I and II, which particularly differ by the existence of a glycine–serine (GS) box in the juxtamembrane domain of type I receptors. Additionally, membrane proteins of the receptor cell are needed for signal transduction and work as coreceptors. Numerous nomenclatures have been used, which supplements to the natural complexity of this system.

- There are seven type I receptors, called activin receptor-like kinase (ALK), from ALK1 to ALK7. The type I receptor of the three TGFβs is distinctive and is named TβR1 or ALK5. The type I receptors of the other members of the TGFβ subfamily are ALK2, ALK4 and ALK7 and those of the BMP subfamily are most commonly ALK1, ALK3 and ALK6; the type I receptor of the AMH is ALK2 or ALK3.
- There are five type II receptors; likewise, the type II receptor of the TGFβs is distinctive; the type II receptors of the other members of the TGFβ subfamily are ACVR2A and ACVR2B; the type II receptors of the BMP subfamily are either the same ACVR2A and ACVR2B or BMPR2; the type II receptor of the AMH is AMHR2.
- All the coreceptors, occasionally named type III receptors, have not all been recognized; for the TGFβs, this is a proteoglycan named β-glycan or TβR3; for the TGFβ subfamily, this is a protein

named cripto; for the BMP subfamily, these are either repulsive guidance molecule A and B or endoglin for BMP9 and 10. The mechanism of action of the coreceptors is not fully stated and may differ according to the receptor complexes and the ligands.

- Some decoy receptors have been recognized, such as the BMP and activin membrane-bound inhibitor homologue protein; they have structural analogy with type I receptors but only a decreased intracytoplasmic portion, short of kinase activity. These molecules are negative regulators of the signaling pathway.

Most of the dimeric factors of the TGFβ and BMP subfamilies bind first to two molecules of type II receptor, which then engages two molecules of type I receptor; this is the opposite for some BMP factors such as BMP2 and 4. In all situations, the active complex is a hexameric complex that has been crystallized for some of them, permitting X-ray diffraction structural studies. The diversity of the probable combinations demonstrates the diversity of the messages that can be transduced by this signaling pathway.

The phosphorylation of type I receptor takes place at the level of the GS box, upstream from the catalytic domain. This phosphorylation stimulates a conformational change which unravels the catalytic site of the type I receptor, which thus can phosphorylate the SMAD proteins. The protein FK506-binding protein can bind the GS box when it is not phosphorylated and thus hinders receptor activation in the absence of the TGFβ signal.

5.3. SIGNAL TRANSDUCTION IN THE TGFB PATHWAY

In distinction to the complexity of the combinatorial pattern of ligands and receptors, there is a relative plainness of the signal transduction pathway (Fig. 5.3). The TGFβ subfamily factors are accountable for the activation by phosphorylation of SMAD2 and SMAD3 transcription factors, while those of the BMP subfamily stimulate the same for SMAD1, SMAD5 and SMAD8. The binding of these 'primary' SMAD proteins (R-SMADs) to SMAD4 (co-SMAD) stimulates their migration to the nucleus where they recognize and bind the promoter sequences of their respective target genes. Two proteins of the same cluster, SMAD6 and SMAD7 (I-SMADs), are competitive inhibitors of R-SMADs at the level of the activated receptors and hinder their phosphorylation. ALK1 can be engaged by TGFβ when expressed at sufficient levels in endothelial cells, resulting in the activation to SMAD1, 5 or 8 as an alternative to SMAD2 or 3 and, thus, to different, even contradictory actions on the target gene promoters.

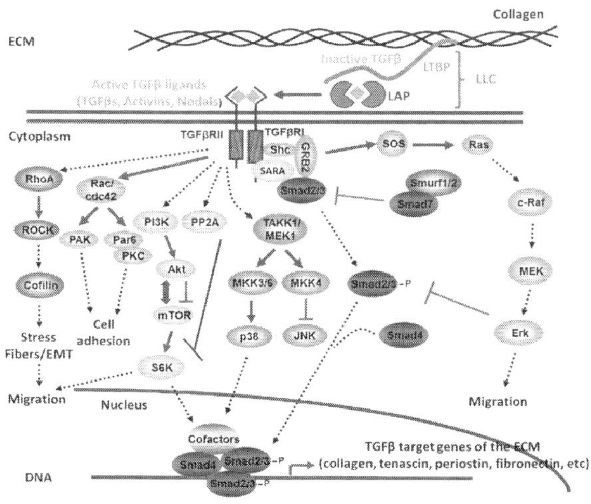

Figure 27 TGFβ signaling pathway (Papageorgis and Stylianopoulos, 2015).

Numerous proteins are implicated in the regulation of SMAD protein phosphorylation by the activated reception complexes; particularly, SMAD anchor for receptor activation engages certain R-SMADs to the activated receptor and seems to be needed for their phosphorylation. The destruction of the reception complexes is guaranteed by endocytosis and proteolysis in the proteasome after the action of an E3 ubiquitin ligase named SMAD ubiquitinylation regulating factor.

The SMAD proteins include two functional domains, MAD homology domain 1 (MH1) and MH2, divided by a linker domain (Fig. 28). The MH1 domain is accountable for DNA binding, but the domain responsible for the transcriptional activity is MH2 and is also accountable for the interactions with the other SMAD proteins and the type I receptor. This domain includes the SSXS sequence at the C-terminal extremity, which is the phosphorylation target of the SMAD proteins by the ALK receptors. This phosphorylation stimulates a conformational change which discharges domain MH1 from domain MH2, activating thus transcriptional activity. The SMAD transcription factors engage transcriptional coactivators and corepressors so, according to the context (i.e. the accessibility of nuclear proteins of the transcriptional machinery), the TGFβ stimulus can stimulate or inhibit the transcription of different genes. Nuclear inhibitors of TGFβ

15. Shi, Y., & Massagué, J. (2003). Mechanisms of TGF-beta signaling from cell membrane to the nucleus. *Cell, 113*, 685-700.
16. Smith, A. L., Robin, T. P, & Ford, H. L. (2012). Molecular pathways: targeting the TGF-β pathway for cancer therapy. *Clinical Cancer Research*, *18*, 4514-4521.
17. Wakefield, L. M., & Hill, C. S. (2013). Beyond TGFβ: roles of other TGFβ superfamily members in cancer. *Nature Reviews Cancer*, *13*, 328-341.
18. Whitman, M., & Raftery, L. (2005). TGFβ signaling at the summit. *Development*, 132, 4205-4210.

6
CHAPTER

CYTOKINES PATHWAY

6.1. CYTOKINES

About 40 cytokines can be distributed in two types according to their general structure and that of their receptors. Type I, characterized by the existence of four α helixes, can be generally subdivided in numerous families:

- Interleukin-2 (IL2) family: IL2, IL4, IL7, IL9, IL15 and IL21.
- IL6 family: IL6, IL11, IL27, LIF (leukaemia-inhibiting factor), ciliary neurotrophic factor (CNTF) and oncostatin M.
- IL3 family: IL3, IL5 and granulocyte-macrophage colony-stimulating factor receptor (GMCSF).
- IL12 family: IL12 and IL23.
- Hormones and other haematopoietic growth factors family: granulocyte colony-stimulating factor (GCSF), erythropoietin (EPO), TPO (thrombopoietin, gene THPO), and leptin (LEP or OBS).

Type II comprises interferons and other IL subfamilies: IL10 family (IL10, IL22, IL26) and IL20 family (IL19, IL20, IL24). The cytokines each have numerous functions and their physiological actions are not restricted to a distinctive tissue but are frequently ubiquitous and pleiotropic. These functions are also frequently redundant, so that numerous cytokines may play the same effect of some target tissues or cells.

6.2. CYTOKINE RECEPTORS AND JAK ACTIVATION

Cytokine receptors are membrane proteins with a single transmembrane domain and an extracellular N-terminal portion. They are characterized by the existence of cysteine residues in the distal part of the extracellular domain and by a motif including two tryptophan and two serine residues (WSXWS) on the proximal part of this domain. The intracellular domain does not contain a catalytic center but several well-conserved domains, particularly box 1, close to the transmembrane domain, characterized by a proline-rich motif of eight amino acids, and box 2, including a group of hydrophobic amino acid residues in the prolongation of the α helix of the transmembrane domain, tailed by a cascade of charged amino acid and tyrosine residues capable of binding a phosphate moiety. The two boxes are implicated in receptor binding to a JAK protein. One of these receptors, CNTFR, has no transmembrane and intracellular domains and is linked to the plasma membrane through a molecule of glycosylphosphatidylinositol. Numerous receptors show soluble variant isoforms produced by alternative splicing: this is the situation in the receptors of IL4, IL5, IL7, IL9, LIF, GCSF, CNTF, GH and PRL. These soluble forms work as cytokine transporters and some of them merely as traps, which bind a cytokine without transmitting a signal.

Cytokine receptors must be dimerized to induce the signaling pathway. The dimers and their ligand represent the 'reception complex' distinctive of the communication to be transmitted. This can be a heterodimerization between a ligand-specific monomer (α type) and a common, ligand family-specific, monomer (β or γ type). The most commonly utilized ligand family-specific receptors are IL3Rβ or βc for the GMCSF family, IL2Rγ or γc for the IL2 family and gp130 or IL6ST (IL6 signal transducer) for the IL6 family. Heterodimerization also takes place for the receptors of type II cytokines of the IL10 and IL20 families, with IL10Rβ and IL20Rβ as common receptors for these ILs, respectively. In distinction, for the hormone (GH, PRL, etc.) receptors and the other haematopoietic cytokine (EPO, TPO) receptors, there is only one type of receptors that are homodimerized upon ligand binding. Receptor homodimerization takes place sequentially: the cytokine binds to a first monomer, which stimulates a conformation change permitting the binding of a second monomer and the communication between the two monomers. Receptor heterodimerization is more complex and may tail numerous types of linkage between the ligand-specific monomer and the common, family-specific, monomer, βc, γc or gp130.

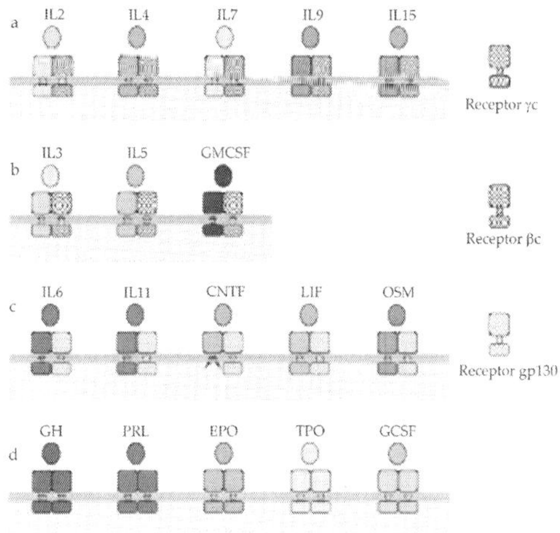

Figure 29 Signal reception complexes of the main type I cytokines (Robert, 2010).

Receptors are constitutively linked to JAK kinases. Receptor dimerization permits the activation of these kinases, which undergo autophosphorylation and then become capable of phosphorylating the receptor at the level of tyrosine residues of box 2. Four JAK proteins have been recognized, named JAK1, JAK2, JAK3 and TYK2. Some receptors are linked to two identical JAK proteins, JAK1 or JAK2, while other receptors are linked to specific JAK proteins: JAK1/JAK2, JAK1/JAK3, JAK1/TYK2 or JAK2/TYK2. No receptor linked to two JAK3 or two TYK2 proteins, or to JAK2/JAK3 or JAK3/TYK2 combinations, has been detected. After autophosphorylation and receptor phosphorylation, the JAK proteins can phosphorylate a cascade of transcription factors of the STAT family, which are engaged to the membrane owing to their SH2 recognition domains of the phosphotyrosine residues of the activated receptor.

The JAK proteins are characterized by the existence of numerous homologous domains JH (JAK homology domains): a JH1 C-terminal domain, possessing the tyrosine kinase activity; a JH2 pseudokinase domain capable of negatively regulating kinase activity; and a four-point-one, ezrin, radixin and moesin N-terminal domain implicated in receptor binding.

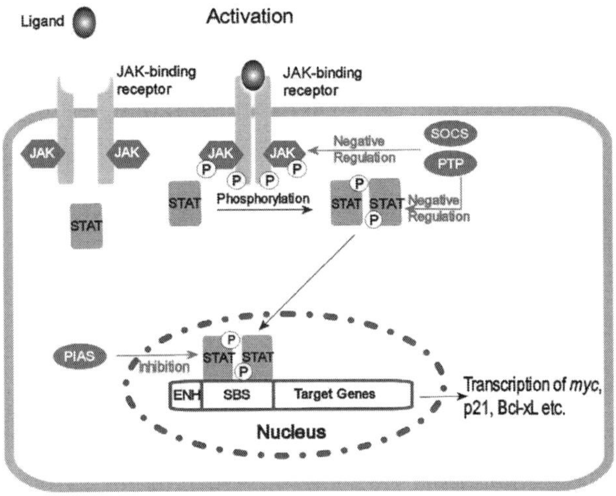

Figure. 30 Cytokine receptor activation via the JAK kinases (Furqan, Mukhi, Lee, & Liu, 2013).

The diverse JAK proteins are redundant but may exert specific roles. JAK1, JAK2 and TYK2 are ubiquitous, while JAK3 is specifically expressed in the haematopoietic tissues and is induced in response to cytokines of the IL2 family. JAK1 is stimulated in reaction to several cytokines, especially of the IL2, IL6 and type II cytokines. JAK2 is distinctive for the reaction to haematopoietic factors (EPO) and hormones. TYK2 is chiefly implicated in type II cytokines response.

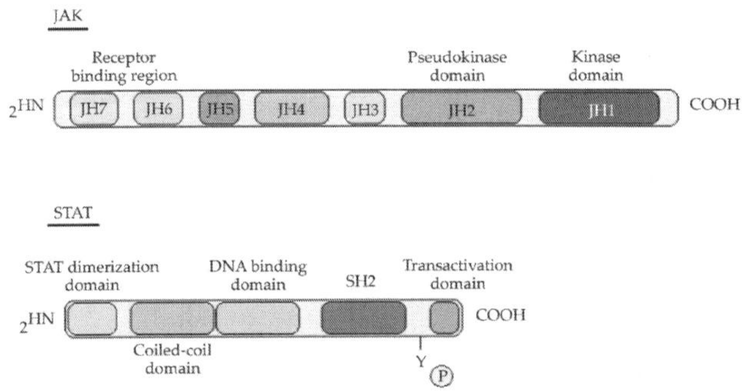

Figure. 31 Structure of the JAK and STAT proteins (Becerra-Díaz, Valderrama-Carvajal, & Terrazas, 2011).

6.3. SIGNAL PROCESSING

The JAK kinases guarantee the phosphorylation of STAT transcription factors. Seven STAT proteins have been detected, denominated STAT1, STAT2, STAT3, STAT4, STAT5A, STAT5B and STAT6. STAT expression is ubiquitous; all have a well-conserved SH2 domain permitting their engagement by the activated receptor on its phosphotyrosine residues and a C-terminal tyrosine residue which is the site of phosphorylation by JAK. This stimulates the formation of phosphorylated STAT protein homodimers or heterodimers, Owing to reciprocal interactions between the SH2 domain of the first one and the phosphotyrosine residue of the second one. These dimers can then move to the nucleus where they bind the promoter sequences of target genes, owing to their DNA-binding domain. One or numerous discrete STAT proteins can be activated in reaction to the binding of a cytokine ligand to its receptor, relying on the cellular context (tissue type, differentiation state, ligand concentration).

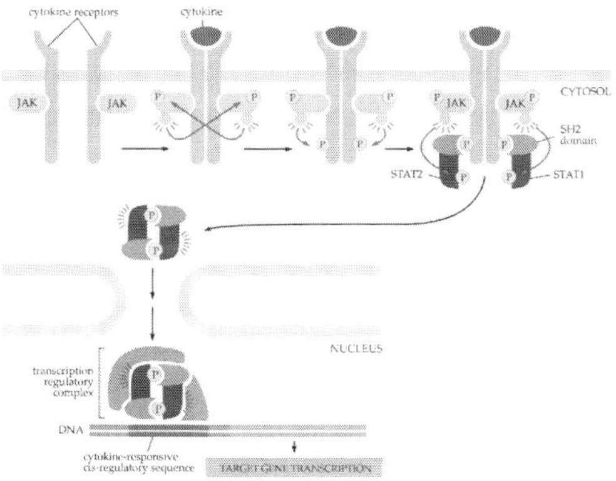

Figure. 32 The JAK–STAT signaling pathway (Cheng, Li, David, Smithgall, Feng, & Lieberman, 2004).

The physiological actions of the STAT proteins are not redundant: STAT1 is relatively specific of the IFNs; STAT3 is selectively stimulated by IL6 family cytokines; STAT4 is stimulated by IL12, STAT6 by IL4 and STAT5 by IL2 and IL3, hormones and haematopoietic growth factors. One of the key regulatory processes of cytokine signal transduction is operated by suppressor of cytokine signaling (SOCS) proteins, also known as STAT-induced STAT inhibitor or JAK-binding proteins. There are eight SOCS proteins: these proteins bear a SH2 domain, which can link to the receptor phosphotyrosine residues and hinder STAT binding to the same activated receptor, and a JAK

6.5. PHARMACOLOGICAL TARGETING

For twenty years, recombinant IL2 (aldesleukin) has been presented in the treatment of advanced renal cancers and malignant melanomas, two cancers that are greatly resistant to chemotherapy and seem to be linked to immune deficiency. IL2 is physiologically generated by CD^{4+} T lymphocytes. It stimulates an augment of lymphocyte proliferation and the clonal expansion of natural killer cells, which constitutively express IL2 receptors. IL2 also induces the production of other factors such as tumor necrosis factor, IL1 and interferon γ.

Contrariwise, an anti-IL2 approach could show interest in the treatment of the cancers for which IL2 may exert a pathogenic role: T-cell leukemia and lymphomas, Hodgkin disease, follicular B-cell lymphomas and hairy cell leukemia. Interferon α is also employed in therapeutics because of its stimulatory activity of natural killer cells and antigen-presenting cells. Moreover, it is prescribed in chronic myeloid leukaemias, hairy cell leukaemias, Kaposi sarcoma and metastatic malignant melanomas. Other interferons (interferons β and γ) are utilized outside the oncology field. The haematological toxicity of high-dose chemotherapy has encouraged the development of therapeutic use of haematopoietic growth factors to hasten the production of progenitor cells. Erythropoietin for the treatment of anaemia, thrombopoietin for the treatment of major platelet deficiencies and GCSF and GMCSF for the treatment of neutropenia have been thus presented in the oncological armamentarium.

Targeting JAK2 activity by tyrosine kinase inhibitors (TKI) lies on the numerous successes that have been got in targeting other TKRs (EGFR) or cytoplasmic tyrosine kinases (mainly BCR–ABL). Ruxolitinib has been approved for the treatment of myeloproliferative neoplasia, without considering the existence of the JAK2 V617F mutation. Other compounds are in progress, but the problem of substrate specificity is certainly one of the most difficult to solve. Numerous JAK3-targeting TKIs are also in progress for the treatment of rheumatoid arthritis, autoimmune diseases and transplant rejection.

Targeting STAT transcription factors, particularly STAT3 and STAT5, is under study. Peptidomimetics targeting the STAT SH2 domains, STAT mRNA antisense oligonucleotides impeding STAT synthesis, decoy oligonucleotides that imitate the DNA-binding site of STAT transcription factors and small molecules recognized by screening represent the numerous tracks that can be tracked.

FURTHER READING

1. Becerra-Díaz, M., Valderrama-Carvajal, H., & Terrazas, L. (2011). Signal Transducers and Activators of Transcription (STAT) Family Members in Helminth Infections. *International Journal of Biological Sciences*, *7*(9), 1371-1381.
2. Benekli, M., Baumann, H., & Wetzler, M. (2009). Targeting signal transducer and activator of transcription signaling pathway in leukemias. *Journal of Clinical Oncology*, *27*, 4422-4432.
3. Cheng, H. Y., Li, P., David, M., Smithgall, T. E., Feng, L., & Lieberman, M. W. (2004). Arsenic inhibition of the JAK-STAT pathway. *Oncogene*, *23*, 3603-3612.
4. Furqan, M., Mukhi, N., Lee, B., & Liu, D. (2013). Dysregulation of JAK-STAT pathway in hematological malignancies and JAK inhibitors for clinical application. *Biomarker Research*, *1*(1), 5.
5. Gadina, M., Hilton, D., Johnston, J. A., Morinobu, A., Lighvani, A., Zhou, Y. J., Visconti, R., & O'Shea, J. J. (2001). Signaling by type I and II cytokine receptors: ten years after. *Current Opinion in Immunology*, *13*, 363-373.
6. Guo, Y., Xu, F., Lu, T., Duan, Z., & Zhang, Z. (2012). Interleukin-6 signaling pathway in targeted therapy for cancer. *Cancer Treatment Reviews*, *38*, 904-910.
7. Haan, C., Kreis, S., Margue, C., & Behrmann, I. (2006). Jaks and cytokine receptors – an intimate relationship. *Biochemical Pharmacology*, *72*, 1538-1546.
8. Ihle JN. Cytokine receptor signalling. Nature. 1995;377:591–4.
9. Ihle, J. N., Witthuhn, B. A., Quelle, F. W., Yamamoto, K., & Silvennoinen, O. (1995). Signaling through the hematopoietic cytokine receptors. *Annual Review of Immunology*, *13*, 369-398.
10. Miklossy, G., Hilliard, T. S., & Turkson, J. (2013). Therapeutic modulators of STAT signalling for human diseases. *Nature Reviews Drug Discovery*, *12*, 611-629.
11. Mullally, A., & Ebert, B. L. (2012). Janus reveals another face: the biologic rationale for targeting Janus kinase 2 in lymphoma. *Journal of Clinical Oncology*, *30*, 4168-4170.
12. Murray, P. J. (2007). The JAK-STAT signaling pathway: input and output integration. Journal of Immunology, 178, 2623-2629.

13. O'Shea, J. J., Gadina, M., & Schreiber, R. D. (2002). Cytokine signaling in 2002: new surprises in the Jak/Stat pathway. *Cell*, *109*, S121-S131.
14. O'Sullivan, L. A., Liongue, C., Lewis, R. S., Stephenson, S. E. M., & Ward, A. C. (2007). Cytokine receptor signaling through the Jak–Stat–Socs pathway in disease. *Molecular Immunology*, *44*, 2497-2506.
15. Pesu, M., Laurence, A., Kishore, N., Zwillich, S. H., Chan, G., & O'Shea, J. J. (2008). Therapeutic targeting of Janus kinases. *Immunological Reviews*, *223*, 132-142.
16. Quintás-Cardama, A., & Verstovsek, S. (2013). Molecular pathways: Jak/STAT pathway: mutations, inhibitors, and resistance. *Clinical Cancer Research*, *19*, 1933-1940.
17. Renauld, J. C. (2003). Class II, cytokine receptors and their ligands: key antiviral and infl ammatory modulators. *Nature Reviews Immunology*, *3*, 667-676.
18. Robert, J. (2010). Textbook of cell signalling in cancer, an educational approach. Paris, France: Springer-Verlag.
19. Santos, C. I., & Costa-Pereira, A. P. (2011). Signal transducers and activators of transcription-from cytokine signaling to cancer biology. *Biochimica et Biophysica Acta*, 1816, 38-49.
20. Tayal, V., & Kalra, B. S. (2008). Cytokines and anti-cytokines as therapeutics – an update. *European Journal of Pharmacology*, *579*, 1-12.
21. Yoshimura, A. (2006). Signal transduction of inflammatory cytokines and tumor development. *Cancer Science 97*, 439-447.
22. Zhang, J., Li, H., Yu, J. P., Wang, S. E., & Ren, X. B. (2012). Role of SOCS1 in tumor progression and therapeutic application. *International Journal of Cancer*, *130*, 1971-1980.

7
CHAPTER

CELL CYCLE CONTROL

7.1 CELL CYCLE PHASES

7.1.1. G1 Phase

G1 phase has the most flexible length among cell cycle phases. If the cells receive an antiproliferative signal, or a signal stimulating terminal differentiation, the cells can delay their advance in the cycle or quit the cycle and enter a quiescent phase, called G0. G1 advancement is controlled by a checkpoint on DNA integrity; once this checkpoint has been passed, the cycle is engaged without any probability of stopping. DNA replication is arranged in G1 through derepression of transcription factors, which allow the transcription of the genes needed for replication. This derepression comprises their release from proteins that retain them and are named pocket proteins. The synthesis of the purine and pyrimidine nucleotides aimed at DNA incorporation needs various enzymes such as dihydrofolate reductase, thymidylate synthase and others; DNA replication itself also needs various enzymes such as different polymerases, primase, helicases, topoisomerases, etc. The promoters of several genes encoding these proteins are activated during the G1 phase.

result in some diseases that are characteristic of DNA fragility such as "ataxia telangiectasia': ATM, ATR and DNAPK. These kinases activate by phosphorylation the checkpoint kinases CHK1 and CHK2; ATM special substrate is CHK1 and ATR special substrate is CHK2. CHK1 and CHK2 can phosphorylate in order the CDC25 phosphatases (CDC25A for CHK1 and CDC25B/C for CHK2), but these phosphorylations constrain CDC25 activity, stopping, therefore, the activation of the cyclin–CDK complexes necessary for cell cycle progression through the G1 → S transition or through the G2→M transition. CHK2 activates, in addition, the inhibitory kinase WEE1, which sustains the inhibition of CDK1 and maintains the cells in G2 phase. Additionally, CHK2 hinders PLK1 and stops thus the realization of the early mitotic events, since PLK1 is implicated in centrosomes formation, chromosomes separation and mitotic spindle set up.

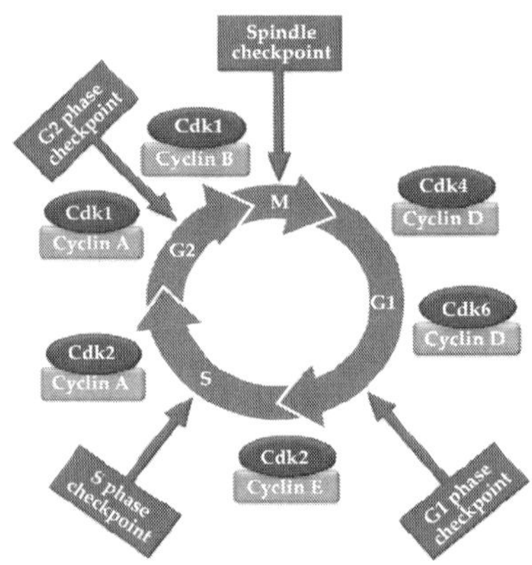

Figure. 39 Cell cycle checkpoints and control of DNA integrity (Spoerri, Oo, Larsen & Haass, 2015).

CHK2 is also capable of activation by phosphorylation the transcription factor p53, which is therefore at the crossroads between the information attained from DNA integrity and the implementation of the pathways targeting the prevention of the propagation of mutagenic DNA damage. p53 is normally present in the cell at very low concentrations, thanks to its binding to its E3 ubiquitin ligase, MDM2 (murine double – minute homologue), which drives it to the proteasome. p53 activation by CHK2-mediated phosphorylation releases it from MDM2 binding and allows it to exert, after homotetramerization, its role of transcription factor. More than hundred genes are transcriptionally activated by p53. This protein, of high importance in oncology as a tumor suppressor, is capable of delaying cell cycle progression in the G1 phase, by stimulating the production of p21 CIP1, a CKI that inhibits cyclin D–CDK4/6 complexes. It is also capable of inducing apoptosis through the transcription of the BH3-only proapoptotic proteins PUMA and NOXA. p53 availability is sustained by the fact that MDM2 is directly inactivated by phosphorylation by ATM. MDM2 is also inactivated by an alternative product of the *CDKN2A* gene, which also encodes a CKI, p16 INK4a. This alternative protein, ARF (alternate reading frame) or p14 ARF, has no structural similarity with p16 INK4a but is implicated in the same tumor suppressor function. p14 ARF production is activated by p53, which hinders, therefore, its own MDM2-driven proteasomal destruction.

7.4. ONCOGENIC CHANGES IN CELL CYCLE REGULATION

The machinery of the cell cycle is vulnerable to be changed in many ways that may share to the loss of its control and to oncogenesis. Various proteins that regulate cell cycle progression can perform as oncogenes products or as tumor suppressors. Without drawing up a list, one can consider as proto-oncogenes the genes encoding CDK4 and CDK6, whose amplification is observed in sarcomas, or cyclin D1, whose normal expression follows the t(11;14) translocation taking place in specific types of lymphomas. CDC25 genes also seem as oncogenic, thanks to their overexpression in numerous cancer types, linked to poor prognosis. MDM2 also works as an oncoprotein: an amplification of the *MDM2* gene, which is located on the same amplicon as CDK4, occurs in liposarcomas and other sarcomas. Without being directly oncogenic, the aurora and the polo-like kinases are commonly overexpressed in cancers.

One can consider as tumor suppressors p16 INK4a and p14 ARF, the gene they share being mutated or invalidated in many leukaemias and solid tumors, as well as the other members of the INK4 or the CIP/KIP families. Loss of function of RB1 is linked to several cancer types. This protein is in an important gatekeeper for cell engagement into DNA replication. Its germinal mutations are specific to the familial form of retinoblastoma. CHK1 and CHK2, as well as the proteins that activate these checkpoint kinases, ATM, ATR and DNAPK, behave as tumor suppressors of the caretaker type and protect the cell against inopportune reproduction when DNA lesions have not yet been illustrated. Last but not least, p53 is the protein whose invalidating mutations are the most common in human cancers; p53 mutations are seen in about 50% of all cancers and usually take place as a late event in cancer evolution. These mutations are extremely diverse and their consequences may differ depending on the alteration of the process of promoter recognition by the changed transcription factor. Among many other effects, p53 mutations hinder it to stimulate cell cycle arrest or apoptosis induction in reaction to DNA damage: the resulting genomic instability is a factor of acceleration of cancer progression.

As previously mentioned, there are two major checkpoints of cell cycle progression; their oncogenic changes allow the cells to perform their proliferation in spite of the presence of DNA lesions, which should result in cell cycle arrest or cell death:

- The G1→S transition, which is specially controlled by the INK4 and the CIP/KIP family; as p21 CIP1 transcription is dependent upon p53, this protein seems as the key factor accountable for this checkpoint.
- The G2→M transition, which is chiefly controlled by WEE1 and MYT1, whose inhibition leads to relieving the mitosis block.

7.5. PHARMACOLOGICAL TARGETING

Cell cycle control seems, thus, as a key element of the regulation of cell proliferation, and its changes share to oncogenesis. Because of their motor role in cell cycle progression and of their pro-oncogenic characteristics, CDKs are viewed as valuable pharmacological targets for cancer treatment, and numerous laboratories are concentrating to the recognition of specific inhibitors of their kinase activity. Screening of natural products has gathered frequent molecules of potential interest, belonging to numerous chemical families: purines, pyridopyrimidines, indirubin, paullones and so forth. However, these molecules are remarkably specific of a single CDK, and the problem is certainly recurrent for serine/threonine kinase inhibitors. In addition, it seems that the numerous CDKs can substitute each other in controlling the consecutive steps of the cell cycle, with the exception of CDK1, which is not expendable. More than a dozen of molecules have gone in clinical trials, such as roscovitine (seliciclib), palbociclib or dinaciclib, which bear a purine moiety with numerous heterocyclic substitutions.

CDC25s also establish potential targets for the recognition of anti-proliferative molecules. Numerous chemical classes of phosphatase inhibitors, especially dicoumarins, are being assessed. The checkpoint kinases CHK1 and CHK2 can also be targeted by kinase inhibitors, with the aim of sensitizing the cells to the anticancer agents that prompt DNA damage: these agents provoke an accumulation of cells in G1 and/or in G2, which limits their effects when the checkpoints of cell cycle progression are working. Similarly, targeting the WEE1 kinase could show to be an appreciated approach, because it firmly controls the G2 checkpoint.

The aurora and polo-like kinases, which control the performance of the crucial phases of mitosis, may also be targeted with suitable kinase inhibitors, which could hinder mitosis eliciting or progression through its successive phases. However, the compounds under study seem to behave as mitotic poisons (such as vinca-alkaloids or taxanes) rather than as targeted therapies. Numerous compounds have gone into clinical trials, either targeting both AURKA and AURKB (tozasertib, danusertib) or showing some specificity for one of the two kinases. Similarly, PLK1 inhibitors have also been developed, such as volasertib or rigosertib. It is continuously difficult to target tumor suppressors, since this would need the restoration of a lost function and not inhibit an excessive catalytic action; however, high-throughput screening of molecules capable of inhibiting p53–MDM2 binding has permitted the identification of a new class of molecules, nutlins that interfere with this association and trigger the proapoptotic functions of p53. They are only active on cells not possessing p53-invalidating mutations. Other types of p53 reactivating approaches are being developed.

The execution of the programs of cell cycle progression can also be hindered during the numerous phases and represent basic targets for cancer treatment, and several drugs targeting the DNA replication or mitosis programs have been used for many years in cancer therapy. In that situation, all proliferating cells are the targets of such drugs, and not only those that have undergone an oncogenic change of the control processes topoisomerase inhibitors and antimetabolites, which block DNA replication (S phase), and spindle poisons that counteract tubulin polymerization (vinca-alkaloids) or hinder microtubule depolymerization (taxanes)—belong to the basic anticancer armamentarium.

28. Robert, J. (2010). Textbook of cell signaling in cancer, an educational approach. Paris, France: Springer-Verlag.
29. Schwartz, G. K., & Shah, M. A. (2005). Targeting the cell cycle: a new approach to cancer therapy. *Journal of Clinical Oncology*, *23*, 9408-9421.
30. Silkworth, W. T., & Cimini, D. (2012). Transient defects of mitotic spindle geometry and chromosome segregation errors. *Cell Division*, *7*(1),19.
31. Sionov, R. V., & Haupt, Y. (1999). The cellular response to p53: the decision between life and death. *Oncogene*, *18*, 6145-6157.
32. Spoerri, L., Oo, Z. Y., Larsen, J., & Haass, N. K. (2014). Cell Cycle Checkpoint and DNA Damage Response Defects as Anticancer Targets: From Molecular Mechanisms to Therapeutic Opportunities. *Stress Response Pathways in Cancer* (pp 29-49). Springerlink.
33. Stark, G. R., (2006). Taylor, W. R. Control of the G2/M transition. *Molecular Biotechnology*, *32*, 227-248.
34. Stolz, A., Ertych, N., & Bastians, H. (2011). Tumor suppressor CHK2: regulator of DNA damage response and mediator of chromosomal stability. *Clinical Cancer Research*, *17*, 401-405.

8
CHAPTER

WNT PATHWAY

8.1. WNT LIGANDS AND THEIR RECEPTORS

About 19 different WNT proteins exist in humans. The first of these genes, WNT1, had been detected at the integration site of a retrovirus inducing mammary tumors in mouse and named Int2, explaining thus a possible role of this gene in normal development and oncogenesis; an ortholog of this gene, named Wingless, had been thereafter detected in Drosophila: the Wnt name come from shorting of Wingless and Integration.

WNT proteins are cysteine-rich glycoproteins liberated by different cell sorts; their genes are organized in many clusters localized on different chromosomes and show high homology. WNT proteins are expressed in a tissue-specific way, but the exact role of each WNT protein is yet poorly stated. WNT ligands are recognized by Frizzled receptors (FZD); there exist 10 FZD receptors in humans, but the combinatorial way between the 19 ligands and the 10 receptors is not fully known. FZD receptor activation needs the existence of a phosphorylable coreceptor, LDL receptor-related protein (LRP5 or LRP6). FZDs are seven-transmembrane domain receptors or G-protein-coupled receptors (GPCRs); the associated activation of an FZD receptor with LRP5/6 coreceptor results in the formation of a complex with a cytoplasmic protein named Dishevelled (DVL), an adapter protein whose activation permits the break of a cytoplasmic complex comprising the essential element of this pathway, β-catenin. Three DVL proteins, DVL1, 2

15. Niehrs, C. (2012). The complex world of WNT receptor signalling. *Nature Reviews Molecular Cell Biology*, *13*, 767-779.
16. Polakis, P. (2007). The many ways of Wnt in cancer. *Current Opinion in Genetics and Development*, *17*, 45-51.
17. Robert, J. (2010). Textbook of cell signalling in cancer, an educational approach. Paris, France: Springer-Verlag.
18. Shi, J., Chi, S., Xue, J., Yang J, Li, F., & Liu, X. (2016). Emerging Role and Therapeutic Implication of Wnt Signaling Pathways in Autoimmune Diseases. *Journal of Immunology Research*, *2016*(105), 1-18.
19. Takebe, N., Harris, P. J., Warren, R. Q., & Ivy, S. P. (2011). Targeting cancer stem cells by inhibiting Wnt, Notch, and Hedgehog pathways. *Nature Reviews Clinical Oncology*, *8*, 97-106.
20. Wang, Y. (2009). Wnt/Planar cell polarity signaling: a new paradigm for cancer therapy. *Molecular Cancer Therapeutics*, *8*, 2103-2109.
21. Wend, P., Holland, J. D., Ziebold, U., & Birchmeier, W. (2010). Wnt signaling in stem and cancer stem cells. *Seminars in Cell and Developmental Biology*, *21*, 855-863.

9 CHAPTER

NOTCH PATHWAY

9.1. DSL LIGANDS

Two families of DSL ligands can detect the Notch receptors in mammals: DLL (delta-like ligands) and JAG (jagged), which interrelate to the Serrate ligands of Drosophila. These ligands are type I transmembrane proteins (i.e. their N-terminal part is in the extracellular space). DSL ligands possess a short intracytoplasmic domain, a single transmembrane domain and a large distinguishing extracellular domain, made of 6–14 tandem repeats of EGF-like domains, a DOS (delta and OSM-11-like) domain absent from DLL3 and DLL4 and a DSL domain. DLL3 and DLL4, short of a DOS domain, need a DLK (delta-like protein homologue) coligand, armed with DOS and EGF-like domains but short of a DSL domain. The concurrent existence of DOS and DSL domains seems essential for the ligand–receptor interaction, the DOS domain being either directly carried by the ligand (for JAG1, JAG2 and DLL1) or by a coligand (for DLL3 and 4). The protein structure of ligands and coligands of the Notch pathway is shown in Figure. 43.

targeting the NRR domain of the Notch receptors and the DLL ligand more specifically implicated in angiogenesis. Demcizumab (anti-DLL4), OMP-52M51 (anti-Notch1) and OMP-59R5 (anti-Notch2) have gone into clinical trials.

FURTHER READING

1. Al-Hussaini, H., Subramanyam, D., Reedijk, M., & Sridhar, S. S. (2011). Notch signaling pathway as a therapeutic target in breast cancer. *Molecular Cancer Therapeutics*, *10*, 9-15.
2. Aster, J. C., & Blacklow, S. C. (2012). Targeting the Notch pathway: twists and turns on the road to rational therapeutics. *Journal of Clinical Oncology*, *30*, 2418-2420.
3. Falix, F. A., Aronson, D. C., Lamers, W. H., & Gaemers, I. C. (2012). Possible roles of DLK1 in the Notch pathway during development and disease. *Biochimica et Biophysica Acta*, *1822*(6), 988-995.
4. Fortini, M. E. (2009). Notch signaling: the core pathway and its posttranslational regulation. *Developmental Cell*, *16*, 633-647.
5. Guruharsha, K. G., Kankel, M. W., & Artavanis-Tsakonas, S. (2012). The Notch signaling system: recent insights into the complexity of a conserved pathway. *Nature Reviews Genetics*, *13*, 654-666.
6. Habas, R., & Dawid, I. (2005). Dishevelled and Wnt signaling: is the nucleus the final frontier? *Journal of Biology*, *4*, 2
7. Kopan, R., & Ilagan, M. X. (2009). The canonical Notch signaling pathway: unfolding the activation mechanism. *Cell*, *137*, 216-233.
8. Nefedova, Y., (2008). Gabrilovich, D. Mechanisms and clinical prospects of Notch inhibitors in the therapy of hematological malignancies. *Drug Resistance Updates*, *11*, 210-218.
9. Qiao, L., & Wong, B. C. (2009). Role of Notch signaling in colorectal cancer. *Carcinogenesis*, *30*, 1979-1986.
10. Ranganathan, P., Weaver, K. L., (2011). Capobianco, A. J. Notch signalling in solid tumours: a little bit of everything but not all the time. *Nature Reviews Cancer*, *11*, 338-351.
11. Rizzo, P., Osipo, C., Foreman, K., Golde, T., Osborne, B., & Miele, L. (2008). Rational targeting of Notch signaling in cancer. *Oncogene*, *27*, 5124-5131.
12. Robert, J. (2010). Textbook of cell signalling in cancer, an educational approach. Paris, France: Springer-Verlag.
13. Shao, H., Huang, Q., & Liu, Z. J. (2012). Targeting Notch signaling for cancer therapeutic intervention. *Advances in Pharmacology*, *65*, 191-234.

14. Yan, M., & Plowman, G. D. (2007). Delta-like 4/Notch signaling and its therapeutic implications. *Clinical Cancer Research*, *13*, 7243-7246.

10
CHAPTER

ION CHANNEL-COUPLED RECEPTORS

10.1. ACTIVATION OF LIGAND-GATED ION CHANNELS

Ionotropic receptors or LGICs are ion channels permitting the transfer through the plasma membrane of excitable cells of anions (Cl^-) or cations (Na^+, K^+, Ca^{2+}). They are dissimilar from voltage-gated ion channels exerting upstream or downstream their activation. The main LGICs are the acetylcholine nicotinic receptor, the serotonin receptor, certain receptors of γ-aminobutyric acid (GABA), glutamic acid and glycine, and ATP receptors named purinergic or P2X receptors. For most of these ligands, other receptors exist, which belong to the GPCR cluster; they are named metabotropic receptors to discriminate them from ionotropic receptors: the acetylcholine muscarinic receptor, amino acid receptors and the second type of purinergic receptors, named P2Y, which also recognize adenylyl nucleotides, distinct from the adenosine receptors P1 or ADORs.

LGICs have a similar common structure, including transmembrane proteins that are assembled together to figure a pore through which ions (anions or cations according to the LGIC) are transferred. There are about 70 protein subunits representing precise functional units, localized particularly at the level of synapses. Some have a pentameric structure (acetylcholine, serotonin, GABA, glycine receptors), others a tetrameric structure (glutamate receptors) or a trimeric structure (purinergic receptors).

11 CHAPTER

APOPTOSIS PROCESS AND REGULATION

11.1. PROTEINS IMPLICATED IN APOPTOSIS

11.1.1. Caspases

Caspases are the group of effector enzymes of apoptosis; these are proteases featured by the existence of a cysteine residue in their active center and by the capability of hydrolyzing polypeptide chains at the level of an aspartic acid residue, which clarifies the name of caspase. Some caspases (class I) are capable of activating signaling molecules through proteolysis, such as interleukin 1 (IL1). Other caspases (class II) are the apoptosis effector caspases and can hydrolyse and eliminate a large variety of cell proteins; caspase 3 (CASP3) is the chief class II caspase. Finally, class III caspases, such as caspase 8 and caspase 9, are the apoptosis initiator caspases, which activate the class II caspases through proteolysis.

Protein activation by proteolysis on special sites is one of the significant mechanisms of post-translational modulation of protein activity. Cell proteins are classically protected from proteolytic enzymes; most of them present therefore under inactive forms (procaspases in the case of caspases), which themselves are activated by certain degree of proteolysis. This needs an initiating event so as to achieve the first proteolytic activity of the cascade. In the case of apoptosis, this event consists in the formation of transient supramolecular edifices, sometimes called platforms, which arrange the procaspases, with the help of adapter proteins, so that they will become able to mutually hydrolyse themselves and thus become activated.

As a common mechanism, the activation of effector caspases needs initiator procaspases, adapter proteins allowing them to organize a platform and the effector procaspases pending for proteolytic activation. In the situation of the intrinsic pathway, the initiator procaspase is procaspase 9, and the activation platform is named the apoptosome; in the condition of extracellular signal-triggered apoptosis, the initiator procaspase is procaspase 8, and the activation platform is named the death-inducing signaling complex (DISC). The activation of a procaspase into a caspase implicates the removal of an N-terminal prodomain, the breakage of a C-terminal domain that stays non-covalently linked to the rest of the molecule and the dimerization of two rearranged caspase monomers (Figure. 49). Initiator and effector procaspases vary by the structure of their N- and C-terminal domains; initiator procaspases harbor in their N-terminal part specific detection domains, which are named caspase recruitment domains (CARDs) or death effector domains (DEDs). All caspases have conserved the same amino acid sequence in their active center (QACXG) and the capability, one activated and dimerized, to split polypeptide chains immediately forward to an aspartic acid residue. Figure. 50 shows the two steps of the activation of an initiator procaspase into a caspase and of an effector procaspase into a caspase as well as the eventual intervention of an activated effector caspase on a target protein.

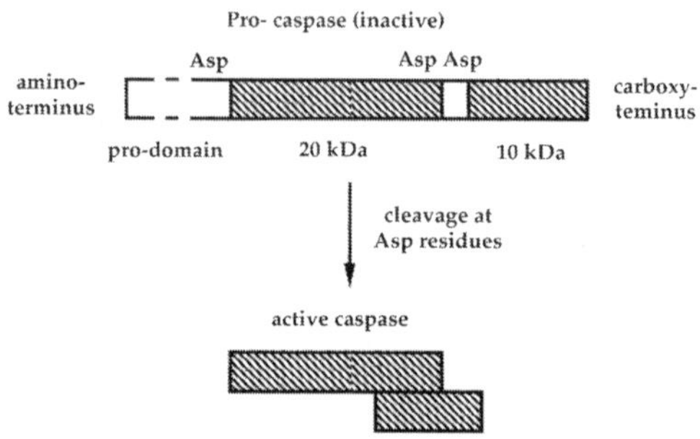

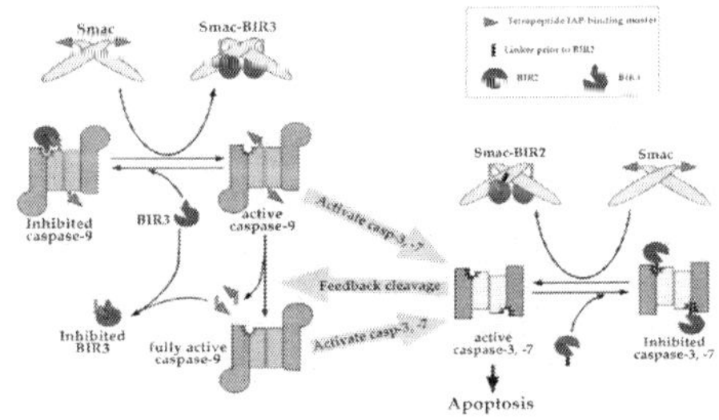

Figure. 49 Caspase structure and activation. (a) General structure of caspases. (b) Caspase activation (Gitendra-Wickremasinghe and Hoffbrand, 1999).

The example represented here is that of caspase 9.

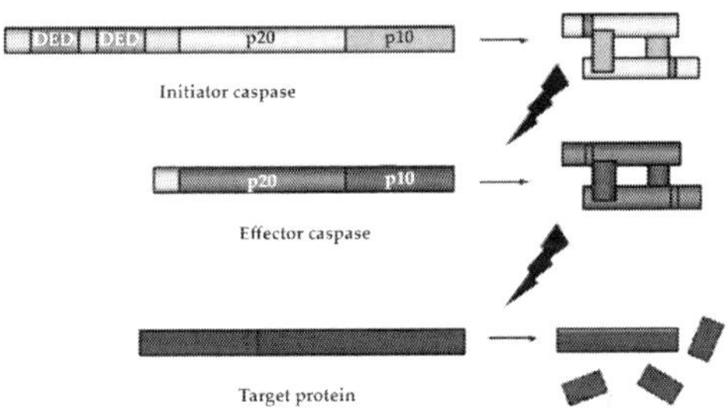

Figure. 50 Caspase proteolytic cascade (Robert, 2010).

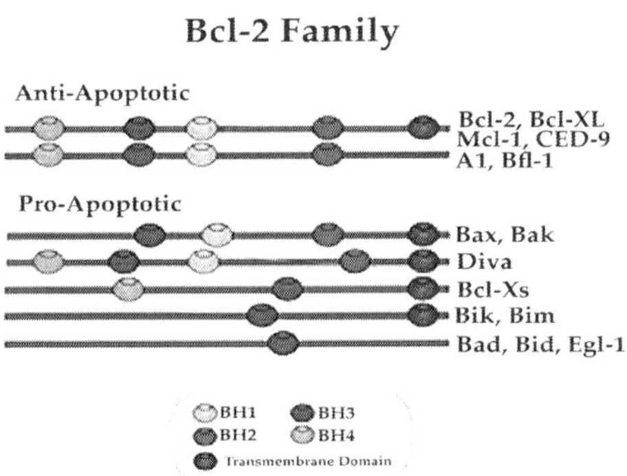

Figure. 51 B-cell lymphoma 2 (BCL2) family proteins (Chao, Korsmeyer, 1998).

reciprocally cleave each other and create active caspase 9, which is then capable of activating procaspase 3 into caspase 3.

Another significant factor capable of exiting out of mitochondria when it becomes permeable is SMAC. This protein is an inhibitor of the proteins that impede caspases (the IAPs). SMAC endorses apoptosis induction confirmed by cytochrome c exit from mitochondria by inhibiting the effector caspases of the cell death programme.

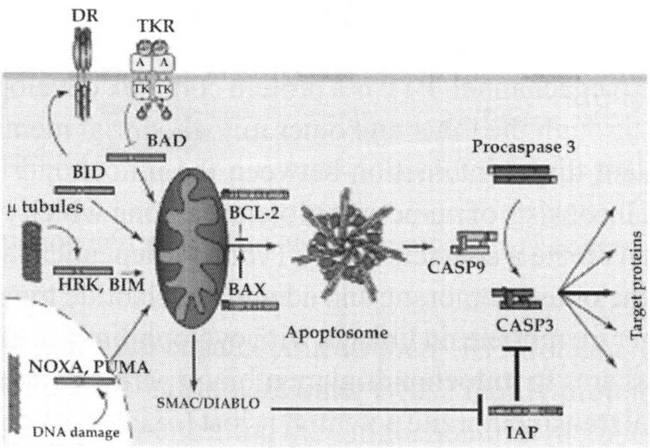

Figure. 55 Intrinsic (mitochondrial) apoptosis pathway (Robert, 2010).

11.3. EXTRINSIC (DEATH RECEPTOR) APOPTOSIS PATHWAY

11.3.1. Activation of the Extrinsic Pathway

The apoptosis-inducing ligands mentioned above, FASL, TNF and TRAIL, are capable of specifically detect their linked receptors FAS, TNFR1 and TRAILR1 and TRAILR2, as well as decoy receptors (Figure. 53). The ligands are basically trimeric transmembrane proteins whose C-terminal part is localized outside the cell; this part can be cleaved to create the soluble forms, called 'death factors'. Ligand–receptor interaction can take place in a juxtacrine pattern when both partners are embedded in the plasma membrane or in a paracrine pattern when the ligand has been cleaved and secreted in the extracellular space. The receptors possess extracellular ligand-binding domains and intracellular domains capable of transducing the death signals. However, the decoy receptors equalize the death signals without transmitting the signal received and show antiapoptotic functions.

The relative expression of true and decoy receptors configures the death message transported by the ligands. Ligand binding to its linked receptor stimulates its trimerization. However, some receptors are already pre-associated as oligomers due to PLAD (pre-ligand assembly domain), so the ligand does not induce trimerization but rather stabilizes the receptor trimers. Receptor activation by trimerization permits the homophilic interaction between their DD and the related DED of the adapter proteins, FADD and TRADD, FADD binding to FAS, TRAILR1 and TRAILR2 and TRADD binding to TNFR1. The DED domains of the adapter proteins are in turn recognized by the related domains of procaspase 8, so a molecular network is constituted, a platform called DISC, which exerts in the locality of the plasma membrane the role exerted by apoptosome in the locality of mitochondria (Figure. 56). Numerous DISCs are linked together at the level of membrane lipid rafts and form macroaggregates.

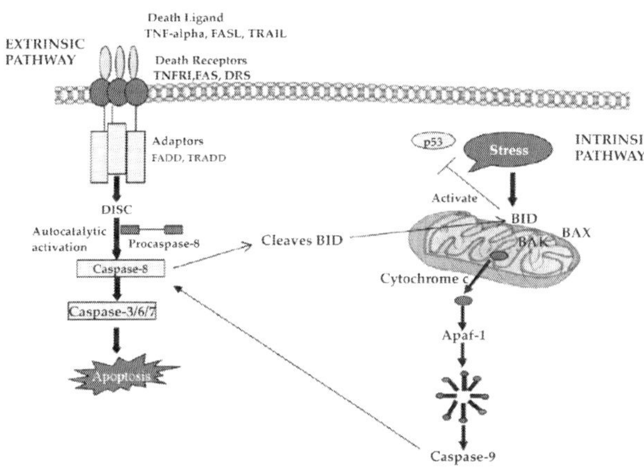

Figure. 56 Extrinsic (death receptors) apoptosis pathway (Rampal, Khanna, Thind, Arora, & Viga, 2012).

11.3.2. Cell Death Induced by the Extrinsic Pathway

The activation of procaspase 8 in the DISC is executed by autocatalytic cleavage, which results in the active form of the molecule. Caspase 8 is then capable of activating procaspase 3 into caspase 3 as already illustrated. In some types of cells, caspase 8 is also capable of activating a BH3-only protein, BID, named t-BID once truncated by caspase 8. As with other BH3-only proteins, t-BID interacts with BCL2 proteins and activates the permeabilization of the mitochondrial outer membrane permitting

FURTHER READING

1. Ashkenazi, A. (2008). Directing cancer cells to self-destruct with pro-apoptotic receptor agonists. *Nature Reviews Drug Discovery*, 7, 1001-1012.
2. Bao, Q., & Shi, Y. (2007). Apoptosome: a platform for the activation of initiator caspases. *Cell Death & Differentiation*, *14*, 56-65.
3. Bredesen, D. E., Mehlen, P., & Rabizadeh, S. (2005). Receptors that mediate cellular dependence. *Cell Death and Differentiation*, *12*, 1031-1043.
4. Chao, D. T., & Korsmeyer, S.J. (1988). BCL-2 family: regulators of cell death. *Annual Review Immunology*, *16*, 395-419.
5. Chaudhary, P. M., Eby, M., Jasmin, A., Bookwalter, A., Murray. J., & Hood, L. (1997). Death receptor 5, a new member of the TNFR family, and DR4 induce FADD-dependent apoptosis and activate the NF-kappa B pathway. *Immunity*, *7*, 821-830.
6. Chipuk, J. E., & Green, D. R. (2008). How do BCL-2 proteins induce mitochondrial outer membrane permeabilization? *Trends in Cell Biology*, *18*, 157-164.
7. Chipuk, J. E., Moldoveanu, T., Llambi, F., Parsons, M. J., & Green, D. R. (2010). The BCL-2 family reunion. *Molecular Cell*, *37*, 299-310.
8. Croft, M., Benedict, C. A., & Ware, C. F. (2013). Clinical targeting of the TNF and TNFR superfamilies. *Nature Reviews Drug Discovery*, *12*, 147-168.
9. Dempsey, P. W., Doyle, S. E., He, J. Q., & Cheng, G. (2003). The signaling adaptors and pathways activated by TNF superfamily. *Cytokine & Growth Factor Reviews*, *14*, 193-209.
10. Ledgerwood, E. C., & Morison, I. (2009). Targeting the Apoptosome for Cancer Therapy. Clinical Cancer Research, 15(2), 420-424.
11. Fesik, S. W. (2005). Promoting apoptosis as a strategy for cancer drug discovery. *Nature Review Cancer*, *5*, 876-885.
12. Fulda, S., & Vucic, D. (2012). Targeting IAP proteins for therapeutic intervention in cancer. *Nature Reviews Drug Discovery*, *11*, 109-124.
13. Wickremasinghe, G. R, & Hoffbrand, V. A. (1999) Biochemical and Genetic Control of Apoptosis: Relevance to Normal Hematopoiesis and Hematological Malignancies. *Blood*, 93, 3587-3600.
14. Johnstone, R. W., Frew, A. J., & Smyth, M. J. (2008). The TRAIL

apoptotic pathway in cancer onset, progression and therapy. *Nature Reviews Cancer*, *8*, 782-798.

15. Kang, M. H., & Reynolds, C. P. (2009). Bcl-2 inhibitors: targeting mitochondrial apoptotic pathways in cancer therapy. *Clinical Cancer Research*, *15*, 1126-1132.
16. Kelly, P. N., & Strasser, A. (2011). The role of Bcl-2 and its pro-survival relatives in tumorigenesis and cancer therapy. *Cell Death & Differentiation*, 18, 1414-1424.
17. Kruyt, F. A. E. (2008). TRAIL and cancer therapy. *Cancer Letter*, *263*, 14-25.
18. LaCasse, E. C., Mahoney, D. J., Cheung, H. H., Plenchette, S., Baird, S., & Korneluk, R. G. (2008). IAP-targeted therapies for cancer. *Oncogene*, *27*, 6252-6275.
19. Lejeune, F. J., & Rüegg, C. (2006). Recombinant human tumor necrosis factor: an efficient agent for cancer treatment. *Bulletin Cancer*, *93*, E90-E100.
20. Lessene, G., Czabotar, P. E., & Colman, P. (2008). BCL-2 family antagonists for cancer therapy. *Nature Reviews Drug Discovery*, *7*, 989-1000.
21. Li, J., & Yuan, J. (2008). Caspases in apoptosis and beyond. *Oncogene*, *27*, 6194-6206.
22. Mahalingam, D., Szegezdi, E., Keane, M., Jong, S., & Samali, A. (2009). TRAIL receptor signalling and modulation: are we on the right TRAIL? *Cancer Treatment Reviews*, *35*, 280-288.
23. Marzo, I., & Naval, J. (2008). Bcl-2 family members as molecular targets in cancer therapy. *Biochemical Pharmacology*, *76*, 939-946.
24. Obexer, P., & Ausserlechner, M. (2014). X-Linked Inhibitor of Apoptosis Protein – A Critical Death Resistance Regulator and Therapeutic Target for Personalized Cancer Therapy. *Frontiers in Oncology*, *4*, 197.
25. Rampal, G., Khanna, N., Thind, T. S., Arora, S., & Viga, A. P. (2012). Role of isothiocyanates as anticancer agents and their contributing molecular and cellular mechanisms. *Medicinal Chemistry & Drug Discovery*, *3*(2), 79-93.
26. Robert, J. (2010). Textbook of cell signalling in cancer, an educational approach. Paris, France: Springer-Verlag.
27. Ryan, B. M., O'Donovan, N., & Duffy, M. J. (2009). Survivin: a new target for anti-cancer therapy. *Cancer Treatment Reviews*, *35*, 553-562.

28. Villa-Morales, M., & Fernandez-Piqueras, J. (2012). Targeting the Fas/FasL signaling pathway in cancer therapy. *Expert Opinion on Therapeutic Targets*, *16*, 85-101.
29. Vogler, M., Dinsdale, D., Dyer, M. J., & Cohen, G. M. (2009). Bcl-2 inhibitors: small molecules with a big impact on cancer therapy. *Cell Death & Differentiation*, *16*, 360-367.
30. Wiezorek, J., Holland, P., & Graves, J. (2010). Death receptor agonists as a targeted therapy for cancer. *Clinical Cancer Research*, *16*, 1701-1708.
31. Yip, K. W., & Reed, J. C. (2008). Bcl-2 family proteins and cancer. *Oncogene*, *27*, 6398-6406.
32. Youle, R. J., & Strasser, A. (2008). The BCL-2 protein family: opposing activities that mediate cell death. *Nature Reviews Molecular Cell Biology*, *9*, 47-59.
33. Yu, J. W., & Shi, Y. FLIP and the death effector domain family. *Oncogene*, *27*, 6216-6227.

12 CHAPTER

HEDGEHOG PATHWAY

12.1. HEDGEHOG LIGANDS

Three Hedgehog (HH) proteins exist in mammals; sonic HH, Indian HH and desert HH. They are first synthesized as a 45 kDa precursor, which undergoes an autocatalytic cleavage liberating a 19 kDa N-terminal peptide implicated in signaling and a 45 kDa C-terminal peptide, which is implicated in cleavage and possesses a catalytic activity of cholesterol transferase. HHG proteins are prone to post-translational modifications that are needed for their activity; on the C-terminal side, they are covalently bound to a cholesterol molecule; on the N-terminal side, a specific palmitoyltransferase, HH acyltransferase or SKI (skinny HH), links them to a palmitic acid moiety at the level of a cysteine residue. When armed with these two hydrophobic ligands, HHG proteins can be introduced in the peripheral phospholipid layer of a lipoprotein and thus form a multimeric complex (Figure. 59).

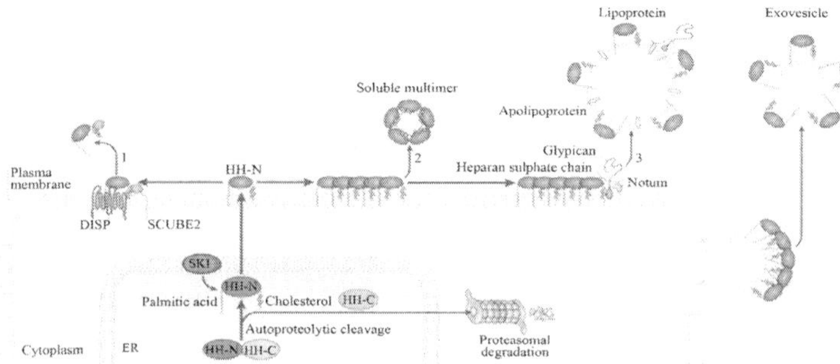

Figure. 59 Hedgehog (HH) protein biogenesis and release (Briscoe and Thérond, 2013).

The mechanism of secretion of the HHG ligands is not fully illustrated. It includes a 12-transmembranem, domain protein named dispatched (DISP), which is essential and behaves as a membrane transporter of HHG proteins. This protein comprises a sterol-binding domain, which should interfere in the processes of recognition and secretion of the HHG proteins.

12.2. PATCHED RECEPTORS AND THEIR ACTIVATION

The HHG ligands with their two lipid substituents are detected on a target cell by a receptor named patched (PTCH). There are two different PTCH proteins, encoded by the genes PTCH1 and PTCH2, which appear to have distinct functions. PTCH belongs to the same transporter family as DISP and also shows 12 transmembrane domains and a cholesterol-binding domain. It also shows two large extracellular domains, which guarantee HHG binding. PTCH is only an intermediary in HHG signal reception, as it transfers the information therefore received to another membrane protein named smoothened, by liberating the constitutive repression played in this protein. The mechanism of this release is not fully illustrated, but it appears that no direct interaction presents between PTCH and SMO. PTCH would only be a transport protein and would control the localization of a small inhibitory or excitatory molecule that keeps to be recognized. In the nonappearance of signal, PTCH would drive an SMO inhibitory molecule from the inside to the outside of the cell; when bound to HHG, PTCH would stop pumping and SMO would then be functional; the contrary would be true if the molecule could contrariwise activate SMO. Some authors assume that the small molecule transported by PTCH and interrelating with SMO could be vitamin D3 or an oxysterol.

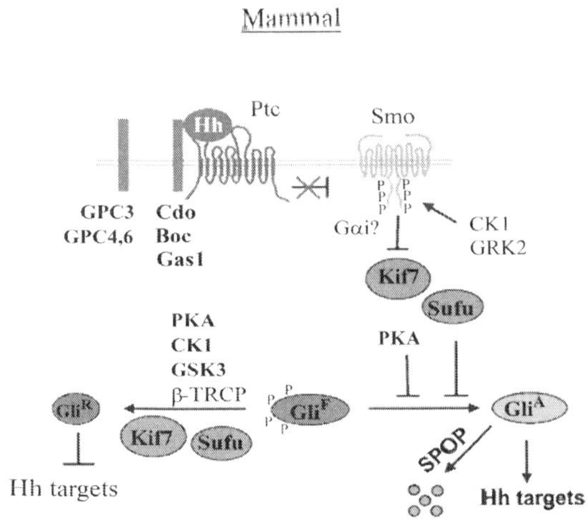

Figure. 60 PTCH receptor information transduction to SMO proteins (Chen and Jiang, 2013).

SMO is a membrane protein with seven transmembrane domains, related to the fifth family of G-protein-coupled receptors as FZD receptors. When PTCH binds an HHG ligand, SMO collects in primary cilia, which are subcellular organelles existing in all vertebrate cells and form protrusions at the surface of the plasma membrane, short of motility and based upon a microtubule cytoskeleton. SMO movement to primary cilia is needed for HH signaling. This migration is supplemented by SMO phosphorylation by GRK2 (GPCR kinase 2), which is thought to be needed for moving. Proteins other than PTCH are capable of binding HHG ligands and modulate the signal brought to the target cell: CDO (cell adhesion molecule-related/ downregulated by oncogenes) and BOC (brother of CDO), which are members of the immunoglobulin superfamily and function as auxiliary receptors to HHG; it is not stated whether or not they are needed for signal transmission. They operate through a glycosaminoglycan (heparan sulfate)-binding site. Another protein, HH-interacting protein (HHIP), would be a decoy receptor for HHG ligands and would thus repress this signaling pathway.

13. Luo, B. H., Carman, C. V., & Springer, T. A. (2007). Structural basis of integrin regulation and signaling. *Annual Review of Immunology*, *25*, 619-647.
14. Mitra, S. K., & Schlaepfer, D. D. (2006). Integrin-regulated FAK-Src signaling in normal and cancer cells. *Current Opinion in Cell Biology*, *18*, 516-523.
15. Rathinam, R., & Alahari, S. K. (2010). Important role of integrins in the cancer biology. *Cancer Metastasis Review*, 29, 223-237.
16. Robert, J. (2010). Textbook of cell signalling in cancer, an educational approach. Paris, France: Springer-Verlag.
17. Stupack, D. G., & Cheresh, D. A. (2002). Get a ligand, get a life: integrins, signaling and cell survival. *Journal of Cell Science*, *115*, 3729-3738.
18. Vicente-Manzanares, M., Choi, C. K., & Horwitz, A. R. (2009). Integrins in cell migration–the actin connection. *Journal of Cell Science*, *122*, 199-206.
19. Ye, F., Kim, C., & Ginsberg, M. (2012). Reconstruction of integrin activation. *Blood*, *119*, 26-33.

14
CHAPTER

SEMAPHORINS

14.1. SEMAPHORINS AND SEMAPHORIN RECEPTORS

Semaphorins represent in vertebrates a group of 20 proteins divided into five classes, from class 3 to class 7. Those belonging to class 3 are released, while the other ones are membrane-bound, either through the presence of a single transmembrane domain (semaphorins of classes 4, 5 and 6) or by anchoring with a glycosylphosphatidylinositol membrane lipid. All semaphorins have a distinctive N-terminal domain named the SEMA domain, 500 amino acids long, which comprises seven-blade β-propeller structure and includes dimerization and receptor-binding sites. They also include other specific domains (Figure. 65). Semaphorins work as homodimers linking two polypeptide chains bound by a disulfide bridge.

13. Sakurai, A., Doçi, C. L., & Gutkind, J. S. (2012). Semaphorin signaling in angiogenesis, lymphangiogenesis and cancer. *Cell Research*, *22*, 23-32.
14. Suzuki, K., Kumanogoh, A., & Kikutani, H. (2008). Semaphorins and their receptors in immune cell interactions. Nature Immunology, 9, 17-23.
15. Takamatsu, H., & Kumanogoh, A. (2012). Diverse roles for semaphorin–plexin signaling in the immune system. *Trends in immunology*, *33*(3), 127-135.
16. Tamagnone, L. (2012). Emerging role of semaphorins as major regulatory signals and potential therapeutic targets in cancer. *Cancer Cell*, *22*, 145-152.
17. Tran, T. S., Kolodkin, A. L., & Bharadwaj, R. (2007). Semaphorin regulation of cellular morphology. *Annual Review of Cell and Developmental Biology*, *23*, 263-292.
18. Zhou, Y., Gunput, R. A., & Pasterkamp, R.J. (2008). Semaphorin signaling: progress made and promises ahead. *Trends Biochemical Sciences*, *33*, 161-170.

15
CHAPTER

NUCLEAR RECEPTOR PATHWAYS

15.1. STRUCTURE AND FUNCTION OF NUCLEAR RECEPTORS

There are 50 nuclear receptors divided into three major classes (Table 3); one of them (NR3 for nuclear receptor 3) encompasses particularly the steroid hormones receptors: estrogens (ERα and ERβ), progesterone (PGR), androgens (AR), glucocorticoids (GR), and mineralocorticoids (MR). Another cluster (NR1) gathers the receptors of thyroid hormones TRα and TRβ, vitamin D (VDR), retinoic acid (RARα, RARβ, and RAR), oxysterols (LXR, liver X receptor), biliary acids (FXR, farnesoid X receptor), fatty acids and eicosanoids (PPARα, PPARβ, and PPARγ, peroxisome proliferator-activated receptor), and numerous xenobiotics (pregnane X receptor [PXR] and constitutive androstane receptor [CAR]). The last cluster (NR2) only includes the accessory receptors of retinoic acid (RXRα, RXRβ, and RXRγ, retinoid X receptors, RXRB, and RXRG).

Table 3 Main nuclear receptors and their ligands (Robert, 2010)

Usual names	Nomenclature	Main lipids
Ma. TRO	**NRIAI.NRIA2**	**Ttryroxim uiio&etryroate**

RAM RARp.RART	**NIMBI. NR 182. NR 1113**	**Alknans mosaic acid**
PPARa. PPARO. MR).	**NR ICI. NR IC2. NR IC3**	**Fay arils. letkotrients. pronaglandins**
ROM. RORp. RORy	**NR 1 Fl. NRIP2. NR 11-3**	**Clualenerot cbolcuarol salchow**
IJCRa.1.):Rp	**RADIIMILD**	**Oxpterols**
MM. ocRo	**NR HU. NRI115**	**Bituy acids**
VDR	**NAM**	**Vitanak D**
PXR.CAR	**NR 1 D. RAID**	**Xenobiatis**
RXRa. Rxitp. RXRy	**N122111. NR2B2. NR2113**	**94.ismomic acid**
ERa and no *(Eral* and *ESR2)*	**NR3A1 et NR3A2**	**OzarafaL unawake**
ERRa. P. 1'	**NMI. NR3B2. NR3113**	**Diedaykabstrol**
CR	**NR3CI**	**Cortical**
MR	**NR3C2**	***Aloacorte***
PR *(POT)*	**NR3C3**	**Pros/stone**
AR	**NR301**	**Teraina**
AHR		**Amami hydrocarlmos (oath)**

Nuclear receptors comprise an N-terminal domain of transcription activation in the nonappearance of ligand (activation function 1 [AF1]), a central domain for DNA binding (DNA-binding domain [DBD]), thanks to two zinc finger domains, a hinge domain, and a C-terminal domain for ligand recognition and binding (ligand-binding domain [LBD]), including the AF2 site for dimerization and ligand-activated transcription (Figure. 68). Relying on the case, nuclear receptors can work as monomers, homodimers, or heterodimers and, in this situation, always with an RXR receptor. They detect on DNA a sequence named hormone-responsive element (HRE). This sequence is either repeated tandemly, constituting two half-sites divided by a variable number of nucleotides, or repeated in a palindromic way, the two half-sites being also divided by a variable number of nucleotides. In both situations, each receptor monomer binds on one of the half-sites.

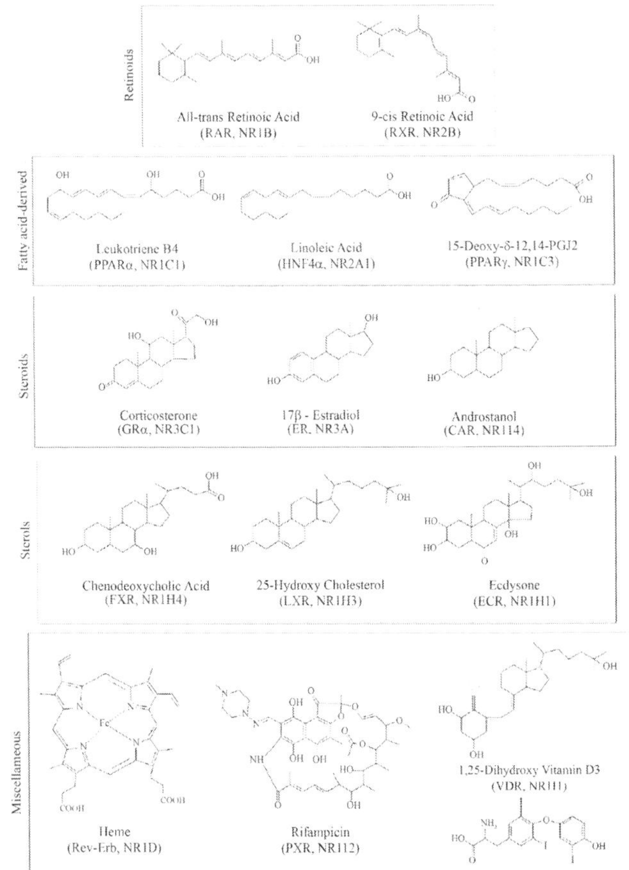

Figure. 67 Structure of the main ligands of some nuclear receptors (Sladek, 2011).

Ligand–receptor binding can take place in the cytoplasm; the translocation of the activated receptor to the nucleus is probable, owing to the unraveling of a nuclear localization sequence. In other situations, the receptor is permanently localized in the nucleus; in the nonappearance of ligand, the LBD is bound to a nuclear receptor corepressor (NCOR); when the LBD has detected and bound a ligand, the transcription activation domain is activated thanks to the intervention of a nuclear receptor coactivator

(NCOA). The function of corepressors and coactivators is to engage histone-modifying enzymes, which control the degree of methylation and acetylation of histones. Histone deacetylation stimulates a compaction of the chromatin that does not permit target gene transcription, while their acetylation improves transcription (Figure. 69). Nuclear receptors can play other effects, independent of their transcription function: some of them are capable, in the cytoplasm, of activating signaling pathways such as the MAP kinases or the PI3 kinase pathways or the pathways linked to Ca^{2+} release in the cytosol.

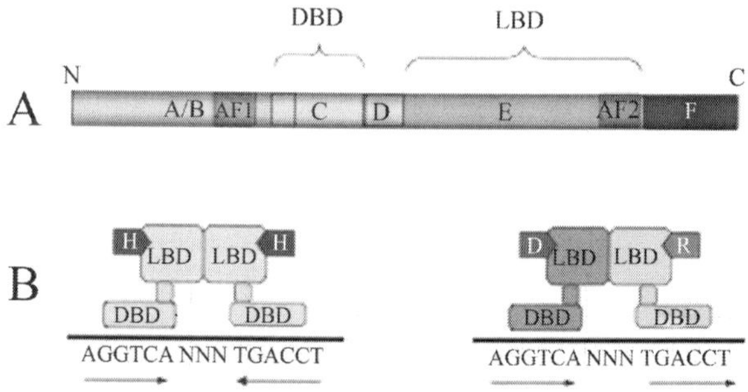

Figure. 68 Structure of nuclear receptors (Robert, 2010).

It is difficult to constitute a list of the genes whose transcription is stimulated by nuclear receptors. Hundreds of genes possess, in their promoter sequence, responsive elements specific for one or many activated receptors. The research of target genes can be achieved in silico, owing to algorithms of identification of these promoter sequences over the whole genome, or by analyzing the gene expression profiles received before and after treatment by a certain ligand. The genes transcribed by a stimulated receptor mirror the characteristics of the related hormones, in the areas of cell metabolism, proliferation, and differentiation. Estrogens and androgens, for example, stimulate a proliferation of the cells which express estrogens and androgen receptors, respectively, which can be utilized by hormone-dependent cancers.

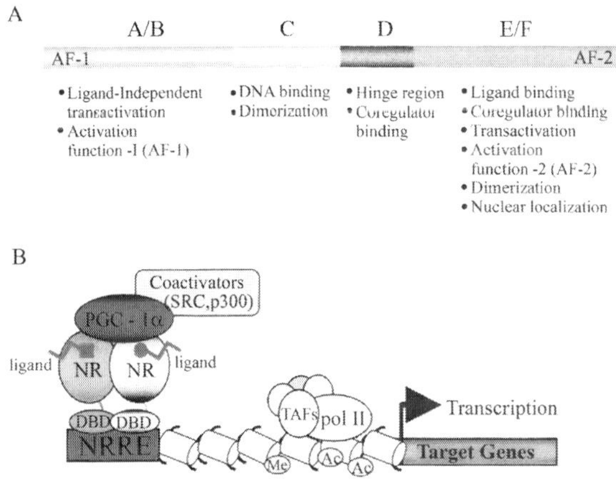

Figure. 69 Transcription activation by nuclear receptors (Huss and Kelly, 2004)

15.2. STEROID HORMONES RECEPTORS

Five clusters of steroid hormones can stimulate nuclear receptors: estrogens, progesterone, androgen, glucocorticoids, and mineralocorticoids. In the nonappearance of ligand, these receptors are linked in the cytoplasm to chaperone heat shock proteins (HSPs) which keep them in an inactive state. Ligand binding stimulates the removal of the HSP, the homodimerization of the receptor, and its translocation into the nucleus. The dimer detects the target sequence HRE and induces the transcription of the corresponding genes.

15.2.1. Estrogens and Progesterone Receptors

The activity of estrogens on breast cancer growth has been stated long time ago since castration was suggested more than one century ago as breast cancer treatment. Estrogens favor the proliferation of tissues expressing ERα, which takes place in 75% of breast cancers. Estrogens, employed as hormone replacement therapy of menopause, have stimulated an increase in breast cancer incidence, which is now declining, as an international warning on their use has been outstretched. The second estrogen receptor, ERβ, is a negative regulator of ERα in normal breast tissue, in regulating its transcriptional activity through heterodimerization, which diminishes the transcription rate.

PPARγ has been predominantly studied in relation with cancer; its implication in oncogenesis is still debatable: it is anti-oncogenic in most models but pro-oncogenic in some situations. Thiazolidinediones, which are PPARγ agonists and are employed in the treatment of type II diabetes, have a protective effect on tumor occurrence in mice. PPARγ mutations have been noticed in colorectal cancers, as well as translocations resulting in fusion proteins with negative dominant effect in thyroid cancers and homozygous mutations in prostate cancers. In distinction, thiazolidinediones may induce tumor evolution of APC-mutated colic adenomas, which must result in coloscopic surveillance of diabetic patients treated with this drug. Indirectly, cyclooxygenase 2 (COX2), which generates potent PPARγ agonists, is implicated in the pro- or anti-oncogenic effects of these nuclear receptors. The combination of COX2 inhibitors and PPARγ agonists could play a synergistic effect against tumor growth.

The oncogenic role of PPARβ/δ is even less obvious than that of PPARγ; its stimulation may have pro-oncogenic penalties (resistance to apoptosis, increase of migratory capabilities) and anti-oncogenic consequences (decline in cell proliferation), agreeing to the tissue or cell context. PPARα relatively seems pro-oncogenic; its pharmacological agonists are fibrates, which stimulate in vitro and in vivo activation of hepatocyte proliferation and repression of apoptosis. These drugs show a long-term liver carcinogenic effect in rodents, but this has not been noticed in humans.

15.7. XENOBIOTICS RECEPTORS

The xenobiotics receptors, PXR, and CAR were seen primarily as orphan receptors as they do not bind any endogenous compound with high affinity. They recognize in fact a large variety of ligands and exert a major role against the deleterious effects of the accumulation of toxic exogenous and endogenous compounds. They are chiefly expressed in the liver, kidney, and intestine. PXR and CAR target the genes deciphering drug metabolizing enzymes, particularly the cytochromes P450, and drug transport proteins, such as P-glycoprotein and other ABC transporters. They exert a key role in the detoxification of xenobiotics and of toxic endogenous products (bilirubin, biliary acids). In oncology, these “xenoreceptors” can modify cancer risk in addition to the activity of anticancer drugs.

As the other receptors of the NR1 family, they heterodimerize with an RXR receptor, bind in the nonappearance of ligand to xenobiotic-responsive elements, and inhibit the transcription of target genes; in the existence of a ligand, this inhibition is eased and transcription can take place. Among the target genes that have been well characterized are cytochrome P450 3A4 (CYP3A4), which guarantees the oxidative metabolism of a large diversity of drugs; UDP-glucuronosyltransferase 1A1 (UGT1A1), which conjugates drugs and endogenous compounds to glucuronic acid; and P-glycoprotein (Pgp, MDR1 for multidrug resistance 1), a transporter implicated in the removal of many xenobiotics out of the organism. Xenoreceptor agonists have not always been recognized precisely; they had been recognized initially as stimulators of cytochromes and ABC transporters: rifampicin, phenobarbital, hyperforin, paclitaxel, ritonavir, and so forth. These compounds work as powerful repressors of the activity of oral drugs because of this induction phenomenon.

PXR and CAR expression is controlled by promoter methylation, which is upper in tumor cells than in normal cells, at least for NR1I2; by micro-RNAs at the level of the 3′ UTR of the genes; and by the negative dominant effect of a truncated variant of the protein, received by alternative splicing. Aryl hydrocarbon receptor is a xenoreceptor not belonging to the nuclear receptor family, but exerting an analogous role as PXR and CAR in xenobiotics metabolism.

It is bound in the cytoplasm to chaperone proteins such as HSP90; when it binds to a ligand, it moves to the nucleus, leaves its chaperone, and binds to an intermediate protein named aryl hydrocarbon nuclear translocator. Once in the nucleus, it binds DNA sequences named dioxin response elements and stimulates the transcription of numerous genes, among which are CYP1A1, CYP1A2, CYP1B1, UGT1A1, GSTP1, ALDH3A1, and NQO1.

FURTHER READING

1. Ali, S., & Coombes, R.C. (2002). Endocrine-responsive breast cancer and strategies for combating resistance. *Nature Reviews Cancer, 2,* 101-112.
2. Altucci, L., Leibowitz, M. D., Ogilvie, K. M., de Lera, A. R., Gronemeyer, H. (2007). RAR and RXR modulation in cancer and metabolic disease. *Nature Reviews Drug Discovery, 6,* 793-810.
3. Aranda, A., Martínez-Iglesias, O., Ruiz-Llorente, L., García-Carpizo, V., Zambrano, A. (2009). Thyroid receptor: Roles in cancer. *Trends in Endocrinology and Metabolism. 20,* 318-324.
4. Chen, Y., Tang, Y., Guo, C., Wang, J., Boral, D., & Nie, D. (2012). Nuclear receptors in the multidrug resistance through the regulation of drug-metabolizing enzymes and drug transporters. *Biochemical Pharmacology, 83,* 1112-1126.
5. Conzen, S. D. (2008). Nuclear receptors and breast cancer. *Molecular Endocrinology, 22,* 2215-2228.
6. Deeb, K. K., Trump, D. L, & Johnson, C. S. (2007). Vitamin D signalling pathways in cancer: Potential for anticancer therapeutics. *Nature Reviews Cancer, 7,* 684-700.
7. Francis, G. A., Fayard, E., Picard, F., & Auwerx, J. (2003). Nuclear receptors and the control of metabolism. *Annual Review of Physiology, 65,* 261-311.
8. Frankfurt, O., & Rosen, S. T. (2004). Mechanisms of glucocorticoid-induced apoptosis in hematologic malignancies: Updates. *Current Opinion in Oncology, 16,* 553-563.
9. González-Sancho, J. M., García, V., Bonilla, F., & Muñoz, A. (2003). Thyroid hormone receptors/THR genes in human cancer. *Cancer Letters, 192,* 121-32.
10. Gronemeyer, H., Gustafsson, J. A., & Laudet, V. (2004). Principles for modulation of the nuclear receptor superfamily. *Nature Reviews Drug Discovery, 3,* 950-964.
11. Handschin, C., & Meyer, U. A. (2003). Induction of drug metabolism: The role of nuclear receptors. *Pharmacological Reviews, 55,* 649-673.
12. Hu, X., & Lazar, M. A. (2000). Transcriptional repression by nuclear hormone receptors. *Trends in Endocrinology and Metabolism, 11,* 6-10.

13. Huss, J. M., & Kelly, D. P. (2004, September 17). Nuclear receptor signaling and cardiac energetics. *Circulation Research*, *95*(6), 568-578.
14. Lee, J. S., Kim, K. I., & Baek, S. H. (2008). Nuclear receptors and coregulators in inflammation and cancer. *Cancer Letters*, *267*, 189-196.
15. Massard, C., & Fizazi, K. (2011). Targeting continued androgen receptor signaling in prostate cancer. *Clinical Cancer Research*, *17*, 3876-3883.
16. Michalik, L., Desvergne, B., & Wahli, W. (2004). Peroxisome-proliferator-activated receptors and cancers: Complex stories. *Nature Reviews Cancer*, *4*, 61-70.
17. O'Malley, B. W., & Kumar, R. (2009). Nuclear receptor coregulators in cancer biology. *Cancer Research*, *69*, 8217-8222.
18. Pascussi, J. M., Gerbal-Chaloin, S., Duret, C., Daujat-Chavanieu, M., Vilarem, M. J., & Maurel, P. (2008). The tangle of nuclear receptors that controls xenobiotic metabolism and transport: Crosstalk and consequences. *Annual Review of Pharmacology and Toxicology*, *48*, 1-32.
19. Robert, J. (2010). *Textbook of cell signalling in cancer: An educational approach*. Paris: Springer-Verlag France.
20. Rosenfeld, M. G., & Glass, C. K. (2001). Coregulator codes of transcriptional regulation by nuclear receptors. *Journal of Biological Chemistry*, *276*, 36865-36868.
21. Ryan, C. J., & Tindall, D. J. (2011). Androgen receptor rediscovered: The new biology and targeting the androgen receptor therapeutically. *Journal of Clinical Oncology*, *29*, 3651-3658.
22. Sladek, F. (2011). What are nuclear receptor ligands? *Molecular and Cellular Endocrinology, 334*(1–2), 3-13.
23. Soprano, D. R., Qin, P., & Soprano, K. J. (2004). Retinoic acid receptors and cancers. *Annual Review of Nutrition*, *24*, 201-221.
24. Thorne, J., & Campbell, M. J. (2008). The vitamin D receptor in cancer. *Proceedings of the Nutrition Society*, *67*, 115-127.
25. Wang, T., Xu, J., Yu, X., Yang, R., & Han, Z. C. (2006). Peroxisome proliferator-activated receptor gamma in malignant diseases. *Critical Reviews in Oncology/Hematology*, *58*, 1-14.

16 CHAPTER

SIGNALING BY OXYGEN AND NITRIC OXIDE

16.1. HYPOXIA

The chief signaling pathway implicated in hypoxia response relies on the activation of a transcription factor, hypoxia-inducible factor (HIF), which stimulates the expression of a large group of genes aimed at relieving this situation. The accessible oxygen amount is by itself the signal allowing HIF activation; oxygen thus seems as a signaling molecule, a messenger stimulating the production of appropriate responses.

16.1.1. Activation And Role of Hypoxia-Inducible Factor

HIF is a heterodimeric complex containing an oxygen-sensitive α subunit and an oxygen-insensitive β subunit. There exist three discrete α and two β subunits. In the existence of sufficient oxygen concentration, numerous amino acids of the α subunit undergo post-translational hydroxylation, which drives HIFα to ubiquitinylation and proteasomal destruction. A decline in the accessibility of oxygen stimulates a decrease of this hydroxylation and a relocalization of the α subunit to the nucleus; in the nucleus, the α subunit can be linked to the β subunit and the complex can induce the transcription of target genes (Figure. 70). There are several signaling pathways which stimulate HIF transcription, particularly the proliferation pathways opened by MAP kinases and PI3 kinase. Several transcription factors, such as MYC one aims at reducing NO generation by NOS inhibitors; there are numerous probable ways, particularly peptides mimicking caveolin structure; the second one aims at producing high intratumor NO concentrations in order to get use of its cytotoxic effect.

16.4. GUANYLYL CYCLASE RECEPTORS

In addition to NO-activated cytoplasmic guanylyl cyclases, there is a small group of guanylyl cyclase receptors which show an original signaling pathway for a cognate small group of peptides. These proteins work as receptors for the natriuretic peptides (NPPs), as atrial natriuretic peptide (ANP), brain natriuretic peptide (BNP, gene NPPB), and C-type natriuretic peptide (CNP, gene NPPC). These factors show natriuretic, diuretic, and vasorelaxant properties and are implicated in blood pressure regulation. ANP and BNP work relatively as endocrine hormones and CNP as a paracrine factor in the heart. These receptors are GCA or GUCY2A for ANP and BNP, and GCB or GUCY2B for CNP. They are phosphorylated in the basal state and experience desensitization upon dephosphorylation of specific sites in their kinase homology domain. A third receptor (NRP3), with very short intracellular domain, chiefly helps as a decoy receptor for clearing extracellular NPPs.

Guanylyl cyclase activators, guanylin and uroguanylin, work as ligands for an intestinal endotoxin receptor with guanylyl cyclase activity, and two other membrane guanylyl cyclases, GUCY2D and GUCY2F, are present in olfactory and photoreceptors, respectively, but are not known to show specific receptor functions. Stimulation of GCC improves colon cancer cell growth arrest, and this protein seems as a tumor suppressor. Agonist activation of GCC has been suggested as a prevention method for colorectal cancer. Downstream guanylyl cyclase receptors, cyclic GMP exerts the same roles as those previously mentioned for cytoplasmic-soluble guanylyl cyclases.

FURTHER READING

1. Brahimi-Horn, C., & Pouysségur, J. (2006). The role of the hypoxia-inducible factor in tumor metabolism growth and invasion. *Bulletin du cancer*, *93*, E73-80.
2. Fang, J., Seki, T., & Maeda, H. (2009). Therapeutic strategies by modulating oxygen stress in cancer and inflammation. *Advanced Drug Delivery Reviews*, *61*, 290-302.
3. Foster, M. W., Hess, D. T., & Stamler, J. S. (2009). Protein S-nitrosylation in health and disease: A current perspective. *Trends in Molecular Medicine*, *15*, 391-404.
4. Francis, S. H., Busch, J. L., Corbin, J. D., & Sibley D. (2010). cGMP-dependent protein kinases and cGMP phosphodiesterases in nitric oxide and cGMP action. *Pharmacological Reviews*, *62*, 525-563.
5. Giaccia, A., Siim, B. G., & Johnson, R. S. (2003). HIF-1 as a target for drug development. *Nature Reviews Drug Discovery, 2*, 803-811.
6. Gorrini, C., Harris, I. S., & Mak, T. W. (2013). Modulation of oxidative stress as an anticancer strategy. *Nature Reviews Drug Discovery*, *12*, 931-947.
7. Grek, C.L., & Tew, K. D. (2010). Redox metabolism and malignancy. *Current Opinion in Pharmacology, 10*, 362-368.
8. Hancock, J. T. (2009). The role of redox mechanisms in cell signaling. *Molecular Biotechnology, 43*, 162-166.
9. Kaelin, W. G, Jr. (2008). The von Hippel-Lindau tumor suppressor protein: O_2 sensing and cancer. *Nature Reviews Cancer*, *8*, 865-873.
10. Kamata, T. (2009). Roles of Nox1 and other Nox isoforms in cancer development. *Cancer Science, 100*, 1382-1388.
11. Keith, B., Johnson, R. S., & Simon, M. C. (2011). HIF1α and HIF2α: Sibling rivalry in hypoxic tumor growth and progression. *Nature Reviews Cancer*, *12*, 9-22.
12. Leiper, J., & Nandi, M. (2011). The therapeutic potential of targeting endogenous inhibitors of nitric oxide synthesis. *Nature Reviews Drug Discovery*, *10*, 277-291.
13. Miyata, T., Takizawa, S., & van Ypersele de Strihou, C. (2011). Hypoxia. 1. Intracellular sensors for oxygen and oxidative stress: novel therapeutic targets. *American Journal of Physiology-Cell Physiology*, *300*, C226-231.

14. Murad, F. (2006). Nitric oxide and cyclic GMP in cell signaling and drug development. *New England Journal of Medicine, 355*, 2003-2011.
15. Ruan, K., Song, G., & Ouyang, G. (2009). Role of hypoxia in the hallmarks of human cancer. *Journal of Cellular Biochemistry, 107*, 1053-1062.
16. Semenza, G. L. (2012). Hypoxia-inducible factors: mediators of cancer progression and targets for cancer therapy. *Trends in Pharmacological Sciences, 33*, 207-214.
17. Shen, C., & Kaelin, W. G., Jr. (2013). The VHL/HIF axis in clear cell renal carcinoma. *Seminars in Cancer Biology, 23*, 18-25.
18. Trachootham, D., Alexandre, J., & Huang, P. (2009). Targeting cancer cells by ROS-mediated mechanisms: A radical therapeutic approach? *Nature Reviews Drug Discovery, 8*, 579-591.
19. Wilson, W. R., & Hay, M. P. (2011). Targeting hypoxia in cancer therapy. *Nature Reviews Cancer, 11*, 393-410.
20. Wu, W. S. (2006). The signaling mechanism of ROS in tumor progression. *Cancer and Metastasis Reviews, 25*, 695-705.
21. Zhang, J., Yang, C., Feldman, M., & Zhuang, Z. (2017, May 23). Vorinostat suppresses hypoxia signaling by modulating nuclear translocation of hypoxia inducible factor 1 alpha. *Oncotarget, 8*(34), 56110-56125.

INDEX

D

E

F